THE VIEW
FROM
GREAT GULL

THE VIEW FROM GREAT GULL

Michael Harwood

Illustrations by
Richard Edes Harrison

E. P. DUTTON & CO., INC. | NEW YORK

LIBRARY OF CONGRESS CATALOGING IN PUBLICATION DATA

Harwood, Michael.
  The view from Great Gull.

  1. Terns. 2. Birds—New York (State)—Great
Gull Island. 3. Great Gull Island, N.Y.—History.
I. Title.
QL696.C46H37   598.3'3   75-37514

10 9 8 7 6 5 4 3 2 1

Published simultaneously in Canada by
Clarke, Irwin & Company Limited, Toronto and Vancouver

ISBN: 0-525-22867-5
Designed by Dorothea von Elbe

For Mary

Acknowledgments

I had the help of many people as I researched and wrote this book. I would particularly like to thank Helen Hays, director of the Great Gull Island Project for the American Museum of Natural History, and all the Gull Islanders past and present who added to my enjoyment of and my knowledge about the island, notably Roger Pasquier, who first introduced me to Great Gull; Ron Franck, Thomas van't Hof, Kenneth C. Parkes, Robert Arbib, Grace Donaldson Cormons, and Harold Heinz, each of whom either supplied information or helped me get it; and Christopher K. McKeever and his family, who did likewise, and also provided kind hospitality to the Harwoods while assisting in our historical researches on Long Island.

I used a number of library collections, including those of the Suffolk County Historical Society in Riverhead, the Suffolk

Museum and Carriage House in Stony Brook, the Niantic Public Library, the New London Public Library, and the library of the United States Coast Guard Academy; I thank the librarians at each of those institutions. I am especially grateful to Elizabeth B. Knox, curator of the New London County Historical Society; to Dorothy King, curator of the Morton Pennypacker Long Island Collection at the East Hampton Public Library; to Joan Jurale of the Wesleyan University Library; to Mrs. William Woodward of the Southold Free Library; to Kenneth W. Rapp, assistant archivist at the United States Military Academy; to John J. Slonaker, research historian at the Army's Military History Research Collection, Carlisle Barracks; to Nancy T. Manson and Sema Gurun at the library of the National Audubon Society; and, as always, to Marguerite B. Hoadley, Carol J. Simms, and Mary M. Fanning of the Gunn Memorial Library in Washington, Connecticut.

More than a dozen individuals at the National Archives in Washington, D.C., were extremely helpful as I sought material on Little Gull Light and Fort Michie; this was surely one of the most pleasant experiences I have had in years of researching in libraries and archives, besides being very productive. I thank everyone there who made it so.

John Wadsworth, Great Gull's intrepid seagoing taximan, offered suggestions and advice. Lawrence Malloy supplied invaluable information about the island's history in this century. J. W. Sanwald generously loaned me the Frederick Chase journals in his possession. Truman Strobridge, historian for the Coast Guard, provided guidance and research materials. Alvin Josephy of American Heritage's book division made available a copy of his article on the Indians of Long Island Sound. Dr. Joe Webb Peoples of Wesleyan University guided me to material about the geology of the Sound. William H. Drury and Ian C. T. Nisbet of the Massachusetts Audubon Society helpfully provided information about resource materials and current research on terns. Joan Alvarez of the public affairs office at the Navy submarine base in Groton assisted my search for information about coastal-defense forts. Clare H. Nunes offered help with Frederick Chase's Latin.

David and Rachel Titus in Middletown, Connecticut; Richard and Jane Plunkett in New York City; and Richard and Patricia Roth in Washington, D.C., were, as usual, ready at a moment's notice to provide shelter and sustenance to a visiting researcher.

Helen Hays read two drafts of the book, and Kenneth Parkes and Richard E. Harrison, one; each made valuable suggestions that found their way into the final product. So did my supportive editor at Dutton, John Macrae, III.

My wife, Mary, to whom this book is dedicated, made many contributions to the work, not the least of which was her careful and critical readings of the manuscript.

I am most grateful to each of the above, and I must now demonstrate that gratitude by absolving them of any blame for such flaws as may be found herein, such flaws being the author's work alone.

M.H.
Washington, Connecticut
August 18, 1975

Foreword

Some time ago my passion for birds led me into an interest in the world's environment. I had for a number of years been devoting my professional energies to the writing of history, but ultimately I came to the conclusion that the environmental crisis had made the writing of history irrelevant; the world was being looted and poisoned by its dominant species, and the time had come to cease reflecting and to man the barricades, in order that there might *be* a human history to write about fifty or a hundred years hence.

Furthermore, it seemed to me that we were then approaching the breaking point of the biosphere so rapidly, and the limitations of the world's resources had become generally apparent so suddenly, that the crisis itself in effect had no meaningful history. It was without precedents, I thought, particularly in that it involved technological elements that were almost wholly new: virtually overnight, man's exploding population, encouraged by modern medicine, had become a burden on the planet; to feed our rapidly escalating needs, we had developed machines to gouge and suck the earth at a terrifying rate and we had invented thousands of new chemical compounds that the planet, geared to changes over geologic time, was simply unequipped to cope with. To be sure, this had not happened spontaneously. But the origins of the crisis, whatever they were, seemed then of far less moment than its symptoms, which had become uncountably varied, acute, and profoundly threatening; one had best be out on the point, fighting to cure the symptoms.

To some degree, I have since changed my mind. In fact, the weaknesses of the nonhistorical approach began to be evident almost as soon as I began to write seriously about the environment. If one is trying to chart a course for the neighbors to follow, one needs to understand how we got into this fix in the first place. A symbol of that reassessment was my instinctive reaction on first visiting Great Gull Island in Long Island Sound a few years ago. Great Gull was the site of a wildlife research station where studies of birds were being made, chiefly in the breeding season. I had done very little scientific work in ornithology, and the chance to do some of it attracted me to the island. But the work was also being carried out in unique surroundings and circumstances that promised rich material, historical and otherwise; I had a hunch that Great Gull might offer possibilities for synthesis.

Late in April of 1974, I became a part-time member of the volunteer research crew at Great Gull. My passion for birds had led me into environmentalism; since then, it had often served as a springboard to more investigation; now I hoped it would help point me through some new doors. It did.

THE VIEW
FROM
GREAT GULL

EAST END   OBSERVATION POSTS   BATTERY   H.Q.
16" GUN PIT   1   2   3   4   PALMER   5   6   7

THE "CAVERN"   SMALL BATTERY   THE MEADOW   BATTERY PASCO   LANDING

GREAT GULL ISLAND *viewed from NW by N*   0   200   400   600 ft.

SITE OF ARMY WHARF. DESTROYED BY GALE NOVEMBER 1950

landing

GREAT GULL ISLAND

40°-12'-07" North, 72°-07'-09" West

TERN NESTING AREAS

Normal tern nesting habitats are scarce on
the island. However, these terns have succeeded in adapting to alien conditions, uti-
lizing exfoliated concrete, grassy edges of
paved surfaces and under the large boulders
of the rip-rapping; but not nesting in high
grass, bayberry or near the headquarters.

☐   paved foundations and walks

┌─┐
└─┘ (dashed)   unpaved foundations

▭   usable buildings

⊙ ▣   cisterns and manholes

⊠   observation posts

   rip-rap

   grass and weeds

   bayberry

   trees

   sand

DORM

DORM

destroyed O.P.

Cartography based on an April 1972 airphoto with
corrections and additions from measurements in
the field.   © 1975, Richard Edes Harrison

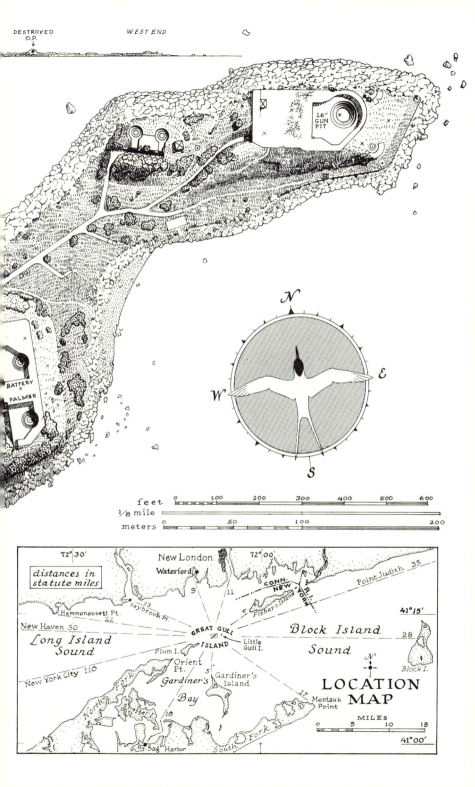

DESTROYED
O.P.
WEST END

16"
GUN
PIT

BATTERY
PALMER

N

W   E

S

feet
⅛ mile
meters

0   100   200   300   400   500   600

0   50   100   200

distances in
statute miles

72°30'   New London   72°00'

Waterford

CONN.
NEW YORK

Point Judith 35

9   11

Fishers Island

13
Saybrook Pt.

Hammonassett Pt.   22

New Haven 50   41°15'

Great Gull
Island

Block Island
Sound   28

Long Island
Sound

Plum I.

Little
Gull I.

New York City 110

Orient
Pt.   5

Gardiner's
Island

Block I.

N

Gardiner's
Bay

North Fork

12   Montauk
Point

LOCATION
MAP

South Fork

Shelter I.

18

MILES

Sag Harbor

0   5   10   15

41°00'

I

April 29: A wet breeze in our faces, and the smell of brine. Our taxi, the fishing-party boat *Sunbeam*, Captain John Wadsworth, is headed south from Waterford, Connecticut, across the eastern end of Long Island Sound toward a small island that lies seven miles south of the New England mainland and seventeen miles northwest of Long Island's Montauk Point. Besides Captain John there are three of us aboard—the ornithologist Helen Hays, president of the Linnaean Society of New York and director of the Great Gull Island Project; Grace Donaldson Cormons, a veteran Gull Island volunteer, who has been sharing in the research for years; and myself, virtually a newcomer. The *Sunbeam* covers some five miles before we can make out anything on the low-lying Great Gull. It was once the site of a coastal defense fort, and from two miles away it looks like a medieval castle island—trim, gray, self-sufficient, with surmounting turrets in the center and the earth of the whole held in a saucer of rock. As we close, the neat gray buildings out at either end become cement blockhouses, and one of them is broken in half, tipped over. The clutch of rectangles near the center resolves into old, boarded-up brick buildings, two slit-windowed cement watchtowers, and a crumbling blockhouse near the beach. The saucer of rock is tumbled granite riprap and seawalls—fortifications laid down by the Army to defend Great Gull against storms.

We have approached across a stretch of water this ruined fort was built to guard. Dead ahead of us, out of sight on the top of a ten-foot bluff, sits the cement emplacement for Battery Pasco, which used to consist of a pair of three-inch guns. A few hundred yards to the east—on our left hand—and on slightly higher ground, another battery pointed a pair of six-inchers north toward New London. Farther to the left, out at the seaward tip of the island, were two ten-inchers, which were later replaced by a single sixteen-inch breech-loading rifle with a barrel about seventy-five feet long; that gun could throw a shell more than twenty miles, well beyond Montauk Point. In the center of the island were two twelve-inch rifles, Battery Palmer, and beyond them, facing south from a bluff on the far shore, stood another six-inch battery.

All the guns are gone, sold for scrap—beaten, one would like to think, into plowshares. The fort was abandoned by the Army more than twenty-five years ago and is now the property of the American Museum of Natural History, which has made the island a wildlife refuge and research station.

Captain John swings the *Sunbeam* around, points her bow toward the Connecticut coast, and backs her down past long, broken rows of pilings—the skeleton of the Army's old main wharf—down to a short, rickety-looking dock, not a yard wide, which is slung between pilings and connected to the stony beach by a scavenged piece of twelve-inch planking. Making a landing on Great Gull can be a tricky business. The tide whips along the shore, and there is very little room for maneuvering in the shoal water; lose headway and you'll be on a rock within twenty seconds. Captain John's predecessor as twice-a-week taximan for the research station, eighty-three-year-old Captain Lawrence Malloy, who also carted supplies for the Army here in World War II, says that even landing at the big dock, farther from the shore, was no picnic. "The steamers landin' there was always more or less troubled," is the way he puts it. "The last dock they had, they built a sort of narrow head on it. That way, they could swing around and kinda head up into the sea better. But when the dock was more or less square"—that is, without any right-angle head at the end—"well, there was only one way that they could land, and most of the boats wouldn't go alongside the dock because they figured there wasn't enough water there."

But the Sound is calm and the current is almost slack this afternoon—"Not Gull Island weather at all," says Captain John rhetorically. "Usually it's rainin', blowin', and rough." The stern line is thrown to a tall, young, black-bearded figure waiting for us on the dock; he makes fast, and Captain John nestles the *Sunbeam*'s bow against the outermost piling. We make fast to that, too, and up onto the narrow dock we heave, piece by piece, jug by jug, carton by carton, a week's worth of food and water and most of a summer's worth of assorted gear for an operation that will eventually involve as many as seventeen people at a time.

In the perpetual ebb and flow of birds along the coast—birds migrating, birds wandering after food—thousands of them pass through this overgrown seventeen-acre patch of sand and rock and rubble each year. Many of them are caught and marked with numbered aluminum leg bands issued by the U.S. Fish and Wildlife Service, and then they are released to continue their journeys. Some of these marked birds will be trapped again eventually, perhaps here, but more likely at banding stations elsewhere; others of them will be found dead. These "recoveries" provide valuable data on the ranges and migration routes and longevity of birds.

Many other birds arrive here to nest—red-winged blackbirds, barn swallows, song sparrows, spotted sandpipers, starlings, yellow throats (lively yellow warblers with black masks on the males), and the terns. All are studied. Most of them, when they can be caught, are given more than just the aluminum bracelets to wear; they are fitted with four-band combinations—the aluminum ring and a colored plastic band on one leg, two colored plastic bands on the other; by varying the colors and the order, the researchers make each bird identifiable from a distance as an individual. (The first people out here each season scour the island for old friends who have made it back for another year.) Thousands of these four-band combinations are made up on strings every winter for marking those old nesters whose combinations have become worn or broken with age, as well as any new nesters that may appear, and the generation to be born during the summer. The marked population will then be watched, and detailed records will be kept of arrival, the competition for territory, courtship, pairing, nesting, the raising of young, and departure.

The birds that get the most attention here are the terns. Fishermen call them "mackerel gulls"; the terns arrive in spring at about the same time the mackerel do, and often when big food fish—such as mackerel, striped bass, and bluefish—are feeding near the surface on schools of baitfish, so will the terns, from above, tipping off the fishermen to where the fishing will be good. It is from the mackerel gulls that Great Gull Island gets its name.

The black-bearded young man who met us at the dock—an eighteen-year-old artist and ornithologist who has been out here by himself since early in the month—tells us as we come ashore that the terns have just the last few days begun to appear over the island. So we are none too early in our own appearance. For the next several weeks they will continue to arrive, until there are about eight thousand of them on Great Gull—two thousand nesting pairs of common terns and nearly as many pairs of roseate terns, along with others of both species that evidently just came along for the trip, being too young or possibly not healthy enough to breed.

They will have traveled a long way to get here, our little mackerel gulls. Common terns spend the winter all along the coasts of South America, down to the Straits of Magellan; a few remain as far north as Florida. The roseate terns put in for the winter in Louisiana, the Bahamas, Cuba, Mexico, and the shores of South America to Brazil and Chile. About the first of April, both species begin to appear in numbers on our southern coasts, and during that month they work their way north, groups of them dropping off as they reach their ancestral breeding territories. For the commons, that may be along the barrier beach of North Carolina and Virginia, or in Long Island Sound, or on islands off Cape Cod and Maine, or in Nova Scotia, or on the shores of the Great Lakes, the Saint Lawrence River, and lakes in the upper Midwest and Saskatchewan. For the roseates, summer home in this hemisphere may be in Venezuela, British Honduras, the Lesser Antilles, Bermuda, the Bahamas, the Tortugas, the Sound, the Cape, or Nova Scotia.

They are, of course, part of the waves and waves of birds that flock northward in April and May—an offshore part, in the East, thousands of gleaming birds, nearly twice the size of robins, joining gulls and cormorants and jaegers and petrels and shorebirds and other species of terns, and pushing ahead of them the sea ducks, geese, loons, and grebes. Millions of birds in motion northward, and among them clouds of swift, angular, gray and white terns, following paths handed down to them, generation upon generation, since the Pleistocene and, through their ancestors, probably since the Paleocene, seventy-five million years ago.

So I enter this story rather late. My first memory of terns I connect with the summer of my twelfth year. My father and his older brother had taken one of my cousins and me cruising on a chartered sailboat out of South Dartmouth, Massachusetts. We sailed across Buzzards Bay, south of the Cape, to the Elizabeth Islands, making port first at Hadley Harbor on Naushon, where everyone cruising south of the Cape put in. The next night, we anchored off some island to the west of Naushon—probably Penikese—and my father and I rowed ashore to dig clams.

As soon as we touched the sand, hordes of black-capped gray and white birds rose screaming from back of the beach and flew at us. Chattering a frantic alarm like soprano machine guns, they hovered over us, dove at our heads and pulled out so close I could feel them pass, then rose, actually flying backwards, to take aim and dive again. Some of them bombed us with their excrement. I found it all pretty unnerving.

Subsequently, I spent a lot of time around water, and I became familiar with the sight of terns, but I was in my twenties before my interest in birds blossomed and I began to pay enough attention to be able to tell the different terns apart. There are a good many tern species found in the East: the gull-billed tern and the black tern, the Forster's, the Arctic, the least, the Caspian, the royal, and the two species that breed on Great Gull, the common and the roseate.

These two birds look very much alike at first glance. Both have gray backs and whitish underparts and incandescent red legs; both have black caps that extend down to the eyes and to the nape of the neck and the top of the bill. Both have forked white tails. But the gray of the common's back is a shade darker than its cousin's; the roseate looks frosty by comparison. The fork in the roseate's tail in breeding plumage appears much deeper than the common tern's, because the roseate's two outermost tail feathers are long streamers—very noticeable. In the spring, the breast of the roseate blooms pink, which is how the bird gets its name. Even their bills are different, the spring common tern having a red bill usually tipped with black, and the spring roseate having a mostly black bill. In the air, although the roseate actually flies the faster, the common tern seems to me the quicker of the two, the more forceful, its flight more driving. The roseate looks pliant, lighter than air, elegant.

They have begun to arrive, but they will not land on Great Gull for days. Strange. As slowly their numbers grow, they will pass over, the commons crying in their strident voices, *Eee-ah, eee-ah, eee-ah, churry-churry-churry, kip, kip,* the roseates calling more softly, *Zaaak, zaaak, zaaak,* and *chivit, chivit.* Pass over, actually avoiding touching this ground; when they rest, they will not rest here, as if they were uncertain, or only mildly interested in what they have presumably flown thousands of miles to reach. But have they? Do these birds we see today "belong" here, after all?

One of the early students of tern behavior, Oliver L. Austin, Jr., noted that in 1932 the terns he was watching on Tern Island, near Chatham, Massachusetts,

> arrived off the Chatham coast, according to the fishermen who make daily trips to the outside cod grounds, on April 20th, and remained off shore as the flocks increased in size, until suddenly, on April 26th, they "came to land" at Tern Island. They flew around the island and rested on the neighboring beaches and sand-bars for two days, and then returned to fish on the outside shoals for another week. They returned to commence their nesting activities on May 3d, but only in about one-third the numbers usually present in the rookery at the height of the season.

A good many mysteries are bound up in that. Dr. Austin apparently believed that those first terns the cod fishermen saw were just hanging around offshore waiting for some biological or environmental signal that would send them all at once streaming toward Tern Island and nowhere else. But there was no proof that they were Tern Island's birds at all. They might just as well have been migrants headed farther north, or even wanderers that had overshot their marks and would finally come to land many miles to the south. No one knew then. No one knows now. From banding terns we *have* learned that they are strongly attracted to the nesting grounds where they were born or where they have raised young before. They often return. But that bond is far from unbreakable. Birds banded on Gull Island are frequently seen nesting on islands off Cape Cod. Great Lakes terns have bred on the Cape. Cape Cod birds have been found nesting here. So the inherited memories of terns may carry recognition signals for more than one ancestral breeding grounds.

On the other hand, there is a good deal of proof that terns sometimes approach their nesting territory in easy stages. Here at Great Gull, for instance, they touch down first on the rocks offshore or on the dock pilings. Then we can tell that at least some of them "belong" here because they are wearing those gaudy sets of bands on their legs, a mark applied to common and roseate terns. After such off-island touchings down, they will land on the island itself, but only briefly. Over a period of weeks, their attachment to the place grows; their stays become longer; they are less and less likely to spook into what the researchers call a "fly-out." Finally they settle in to breed.

However, that's not the only way they arrive, at least not at other breeding grounds. Instead of the apparently leisurely collecting observed at Great Gull—a few birds more in the neighborhood each day—the terns will sometimes appear in strings of small flocks, most of the birds in a colony making their landfall in the same day or two. Other years they will appear over their territory in one great flock, flying very high, and dash directly down to land on the breeding grounds, without any shilly-shallying; here, it seems, only the late arrivals will do that. Furthermore, no one in the Great Gull Island Project has ever heard of local fishermen hereabouts seeing crowds of terns gathering, say, off the eastern end of Long Island for a week before terns are seen over Great Gull. There are many more questions than answers about the way the terns behave.

We stand and crane our necks as they pass over, in ones and twos and threes; we watch for a *sign*. If only we could get a good look at their legs—but you can't see bands on high-flying birds. We nibble at the edges of the unknown, and mark down in notebooks the times they appear, how many there are, which direction they are headed in, and what they are doing as they fly.

Ornithologists have been studying tern behavior in detail for a bit more than a century—not long, compared, say, to the centuries man devoted to the alchemical study of gold, how to make it from scratch. An interesting characteristic of this tern research—indeed, a *crucial* characteristic—is the frequent link between the research and a desire to protect the birds. That's not always the case in bird study, but people began to get interested in terns at about the same time man began to do away with them in wholesale lots. William Brewster, the father of tern research in North America, studied roseate and common and Arctic terns from 1870 to 1874, mostly on the island of Muskeget, which lies between Martha's Vineyard and Nantucket; in 1879 he published an article, "The Terns of the New England Coast," which drew heavily on his research at Muskeget. And in the same breath Brewster, who was one of America's leading ornithologists, took the opportunity to protest the fishermen's persistent thievery of terns' eggs and the "sporting" slaughter of terns by yachtsmen who happened ashore on the nesting grounds.

> Were it not for man,—who, alas! must be ranked as the greatest of all destroyers,—the Terns would here [on Muskeget] find an asylum sufficiently secure from all foes. But season after season the poor birds are daily robbed of their eggs by the fishermen, while frequent yachting parties invade their stronghold and shoot them by hundreds, either in wanton sport or for their wings, which are presented to fair com-

panions. Then the graceful vessel spreads her snowy sails and glides blithely away through the summer seas. All is gayety and merriment on board, but among the barren sand-hills, fast fading in the distance, many a poor bird is seeking its missing mate; many a downy little orphan is crying for the food its dead mother can no longer supply; many a pretty speckled egg lies cold and deserted. Buzzing flies settle upon the bloody bodies, and the tender young pine away and die. A graceful pearl-tinted wing surmounts a jaunty hat for a brief season, and then is cast aside, and [Muskeget] lies forgotten, with the bones of the mother and her offspring bleaching on the white sand. This is no fancy sketch; all over the world the sad destruction goes on. It is indeed the price of blood that is paid for nodding plumes.

And not just the price of blood—a fact that would lead soon to a crisis for the terns.

A Long Island newspaper reported in the 1880s that "within a year or two past the gull shooters of this and neighboring localities have made frequent visits [to Gull Island] and had rare sport in slaughtering numerous of these marine birds. Their plumage is beautiful and prepared for market they find a ready and remunerative sale." Later, Frank M. Chapman, editor of the protectionist magazine *Bird-Lore*, wrote of "hundreds of thousands" of terns being killed on the Atlantic coast. "Cobb's Island, on the coast of Virginia," he said, "is credited with having supplied forty thousand in a single season, and . . . one of the killers recently confessed to me that he knew of fourteen hundred being killed in a day." The hunters would gather up the dead birds, clean the blood off, and pack them in barrels. Chapman said that to dry the freshly washed skins before they were packed, the hunters threw them into a barrel of plaster, "which was rolled up and down the beach until the moisture was absorbed from their plumage. A Long Island taxidermist used a patent churn for this purpose."

The next important student of terns appeared when this commercial shooting for the plume and taxidermy trades was at its peak. He was George Henry Mackay, an old-time gunner-*cum*-ornithologist, who until 1894 was considered an expert chiefly on the habits of shorebirds and sea ducks, seen from the perspective of a hunter looking for sport and table meat. Mackay knew Muskeget, Brewster's study island; he knew the terns and took an interest in protecting them; when the Commonwealth of Massachusetts passed a law forbidding egging in tern colonies, and people in Nantucket and vicinity tried to have the law repealed, Mackay helped defeat the "objectionable petition" and then kept an eye on the Muskeget colony. In the process, he became devoted to terns. During the next five years a steady stream of notes and papers about the terns of Muskeget and Penikese flowed from his pen into the pages of a major American ornithological journal, *The Auk*.

The plume hunting was finally outlawed, and the tern colonies quickly began to recover. But as they did, the terns found themselves increasingly in competition with man for space. Marshes and beaches and islands were taken over for uses that pushed the terns out. Just as this unequal contest was getting fully under way, Drs. Oliver L. Austin, Sr. and Jr., began studying the terns of the outer Cape, covering a much wider geographical area than had been worked previously. And before long they found themselves engaged in protecting the terns from human vandals and animal predators, and in general increasing the terns' breeding success to make up for their diminishing turf.

Rats, for example, are among the most destructive enemies of nesting terns, and tern colonies are often plagued by them. In several of the colonies the Austins were observing rats stole eggs by the hundreds, killed chicks and even parent terns, and terrorized the nesting birds. So the Austins became exterminators, always on the lookout for suspicious-looking burrows and ready to pour in the rat poison at the first sign of trouble.

They also dealt with another common problem in tern nesting areas. Austin the younger wrote in 1932:

> The past four years have witnessed a steady increase in the amount of ground covered by beach-grass on Tern Island and a steady decrease in the patches of open sand. Not only has the grass encroached on the open areas, it has also increased remarkably in height and thickness, no doubt encouraged by the heavy fertilization it receives each summer from guano, dead fish, and dead birds.

He suggested that the vegetation was crowding out the terns—more so the commons than the roseates, which prefer grassy areas—and cutting down on the breeding success of the birds. Experiments bore this out, so two years later a crew went onto Tern Island and pulled out grass by hand, clearing a third of the island that way tern farming. By the 1940s, they were even *plowing* on Tern Island to keep the breeding territory open.

The same mixing of concerns is shown in the current major tern studies in the Northeast—especially here in the Great Gull Island Project and those conducted by the research staff of the Massachusetts Audubon Society from the Cape north. The Massachusetts Audubon scientists have been controlling gull populations around their tern nesting sites, because garbage dumps have caused a population explosion among the gulls; gulls feed not only on the garbage but also on tern chicks and push the terns out of nesting territory. Ian Nisbet of the Massachusetts Audubon staff has even been building little shade boxes for tern chicks, to shelter them from the midsummer sun. On Great Gull we have nothing so fancy as those boxes, but little man-made shelters of flotsam and stone and the Army's nineteenth-century bricks are scattered throughout the nesting areas here. And Gull Island's researchers also follow the Austins' early example, pulling out vegetation by hand: the stone and asphalt and rubble on the island would make short work of a plow.

Since the war, still another threat to the terns has appeared—the heavy pollution of coastal waters with various toxic compounds, oil spilled offshore and in harbors, poisons spread on salt marshes to control insects and chemical wastes poured out of riverside factories, DDT and the like dusted on farms and on the ornamental trees along town streets, which poisons are then washed into the rivers by rain, and thus to the sea. About these pervasive chemicals the researchers cannot do much but shout warnings; the Great Gull Island Project received much of what general fame it has outside the scientific community from the revelation a few years ago that the terns of Great Gull were beginning to show problems that looked suspiciously like those that had already been noticed in bigger birds that feed along shorelines, including the osprey, the peregrine falcon, the bald eagle, the cormorant, and the black-crowned night heron. Deformed tern chicks were being produced, including one sad specimen with four legs; chicks that seemed all right when they

hatched later dropped a lot of their flight feathers before they were old enough to use them. Then the terns began laying eggs that had extremely thin shells or no shells at all, a classic symptom that had previously showed up in the bigger birds and had been blamed on toxic chemicals swallowed in food, chemicals that interfered in some way with the reproductive cycle. Since that time, the symptoms have diminished markedly in the terns on Great Gull, and it may be no coincidence that more care is being taken these days with the use and disposal of the suspected chemicals.

But we are still cutting down our co-planetarians the terns. Ian Nisbet summed up the situation a few years ago: since the 1930s, by his estimate, the population of common terns breeding from New York City northeast to the eastern shores of Maine has been more than halved and stands at about twenty-two thousand breeding pairs; the roseates have remained almost stable, but they number at most only forty-five hundred pairs—close to half of them in the one colony at Great Gull. This is a far cry from what the figures must have been before the circle began to close on these birds, probably much less than half. And, except for the plume hunters, we have produced this result unintentionally, as a side effect of responding to our own needs. But even unintentionally, carelessly, we are too much for them.

Environmentalists, naturalists, and conservationists are often asked to defend their concerns for the well-being of other species of living things besides man. The question is, *What good are they?*—meaning, of course, what good are they *to man*, because if one asked what good they were to the planet, one could just as legitimately ask the same question about mankind. The question often renders the defender of wild things speechless with frustration and rage, or it elicits a defense based on aesthetics, which cuts no ice with the materialistic questioner.

Frank Chapman, trying to answer the question in 1903, remarked dryly that a plume hunter would say that terns "were worth about ten cents each for hat trimmings," while a fisherman would say that "their eggs made excellent omelets; and each has done his best—the one to lay all Terns on the Altar of Fashion, the other to see that none of their eggs escaped the frying pan."

Perhaps the questioner would not be satisfied with any other kind of answer he might get, but if he were, it would be because he was willing to listen to the cosmic view. "Cosmic view" is not the environmentalist's specious synonym for the word "aesthetics"; aesthetics are part of it, to be sure, but in this context they include something deeper than prettiness or decoration of the landscape. These are matters of the human spirit, of connectedness to life, of the complexities of the planet. And practical matters, material matters, are also involved, at a variety of levels.

One must be leisurely, presenting the cosmic view. I think first of Edward Howe Forbush, writing in his *Birds of Massachusetts and Other New England States*. Forbush described a morning in June, 1908, at the outer bend of Cape Cod, where he saw a huge crowd of terns on an island offshore and decided to row out and investigate. A southwest storm had just passed the Cape, the surf was heavy, and as Forbush was landing on the island, his skiff was swamped and the oars were swept away. He managed to drag his boat up on the beach, but without the oars he was literally stranded, so he sat down in the bottom of the skiff to wait for someone to rescue him. While he waited he watched the terns, knowing that they might well provide the means for his resuce.

Not far off the beach where he sat, some of the birds soon discovered a school of baitfish, which had been pushed into the terns' view by a school of larger fish attacking from below. The small fry milled on the surface, perhaps even broke water in their frantic attempts to escape. As the first terns spotted the action, dove, screaming excitedly, and snatched at the fry with their beaks, the rest of the colony saw, heard, and hurried to join the feast. Forbush wrote:

> To see the terns thus fishing is a sight to stir the blood. High in the sunlight they hover above the surging sea. Below the blue waves roar on, to break in foam on the yellow sand. The whirling, screaming, light-winged birds, strongly contrasted against the smoky murk to seaward, alternately climb the air and plunge like plummets straight down into the waves—rising again and again, fluttering, poising, screaming, striking. So now like birds gone mad the terns flashed from sky to sea. It fairly rained birds; hundreds of them were shooting down into the angry waves. . . . This was the sight for which I had been waiting. The birds had given the signal to the fishermen to come out from shore. Soon three dories with their adventurous crews had passed out toward the foaming bar, and the men were dropping their lines near where the fishing birds were thickest. I had only to wait for a returning fisherman to take me off.

That is the proper frame for the cosmic view: the birds are set in their relationship to the sea, the weather, the fish, and man.

Take another instance. One fascinating characteristic of terns is their homing abilities. However far south they go in the fall, in the spring, having flown thousands of miles since they left, many of them make their nests again not only on the same island where they nested the previous year, but in the same general area there, and often on the same spot on the island. Even in the fog, they are unerring navigators. The "local boatmen," remarked George Mackay, "can usually tell under such conditions the bearing of Muskeget Island by watching the course of these birds." The tern flying past the fishermen's boat in fog is headed straight for the ternery if it is carrying a fish in its bill.

The habit of gathering where the big fish are feeding and the other guide services so naturally provided are "uses" of the tern, I admit. But they were not enough to insure their protection in the 1890s, even by the fishermen and yachtsmen who used them as guides; other uses competed. That's the problem. What's useful to me is not necessarily useful to you; in fact, it may not even seem useful to *me*, tomorrow. So a materialistic answer to the materialistic question is not very valuable or lasting. At least, not yet. What we are looking at when we see a cloud of terns diving into a school of baitfish, and what we are looking at when we discover a tern scraping out its nest in the sand within inches of the spot where it made its scrape last year—or later, five miles from Great Gull, when we find it headed home with food for the chicks in that scrape, through fog so thick we cannot see more than two boat-lengths in any direction—is the exceedingly complicated, even miraculous relationships between the things of this earth. We are only *beginning* to know something about these relationships and how they work. At best, we are in a Galilean stage in our understanding of the biosphere. Unfortunately, we are also in a fair way to destroy connectors in those relationships before we know what they are, *because* we don't know and consequently don't care.

We have gained some understanding that there is a great deal of flexibility in our version of a living planet; the biosphere has *give* to it, it has marvelous powers of recuperation. (It better had.) But we do not know how much. In fact, we have no *idea* how much. We appear to be willing to discover where the breaking point is by passing it at full speed. *What good are they?* is a question born of studied ignorance. We have not done much homework, and we are still not doing much homework; our short-term profit-and-loss approach to life tends to slow down such effort. We do have an inkling that variety is not only the spice of life on the planet, but essential to it: "Variety is nature's grand tactic of survival," as one of *Time*'s anonymous reporters wrote a few years ago. *Every* species has value. We have discovered hints of this, among other ways, by eliminating variety in the growing of cash crops, and then observing that one-crop farming over hundreds of acres makes the crops vulnerable to massive plagues, rusts, molds, insects of all sorts—economic disasters that are never so terrible if the crops are varied. We have discovered it by killing off whole races of major predators, and then seeing the former prey, unchecked, multiply beyond the ability of the land to sustain them.

Some of the future answers to the value of these our co-planetarians may seem purely materialistic, on a scale we would recognize now. Research on animals and plants is certain to have innumerable direct applications to improvement of the welfare of our own species, as past research has done. And quite likely, as with many such results of past research, much of what we learn will be unexpected. Benefits will spring from the most unlikely sources. What *good*, for example, is the stinging jellyfish, whose touch is painful and occasionally even fatal to summer swimmers? Resort owners would like to do away with these "pests," have them eradicated from the face of the earth. Recently, however, a Princeton University researcher discovered in the jellyfish a rare chemical substance useful in detecting parathyroid problems in humans.

But future knowledge is certain to have material benefits that go far deeper than that. As we unravel the conundrums of interrelationships, of the fantastically complex web of living, reproducing, hunting, dead, and decomposing things, our species will win insights unimaginable now—insights not only purely practical (how better to grow and store food, to communicate, to find our way in the dark and the fog, how better to prevent war), but aesthetic and spiritual, as well. Simply put, man will feel more at home on his only planet, knowing better how it works and where he fits in.

This view of the value of the things of this earth is clearly a projection well forward into the future. With it we cherish our co-planetarians not only for what they can provide us in our lifetimes but also for what they can provide our children and their children and their children's children. It is an expression of commitment to improve the lives and the understanding of infinite generations of our species.

And the taproot of this commitment is our acceptance of life as a wonder worth passing on. Do we accept life that way now? I think not. Worth passing on to our own seed, perhaps, as extensions of ourselves, as explanation and justification for our own existence. But as for the terns—*what good are they?* And most of our resources go to fulfill goals that can be realized within individual human lifetimes, or less. So we are, at heart, careless of life in the large sense and careless of its meaning. Perhaps this is so because we are appalled by the idea of death. In any case, we persistently reveal ourselves as destructive of life, destructive in fact of self. Despite ominous signs of world overpopulation and starvation and rapidly diminishing resources, we continue to reproduce our species as if our tribes depended on it. We are caught up in our own territorializing. Certainly that is natural, instinctive; the terns do it; and it lies at the heart of our tragedy. We spend hundreds of billions of dollars a year for national defenses—territorializing epitomized. We can think and feel and reason, but we are still trapped in our past.

Walter Sullivan, writing from Sochi in the Soviet Union, October, 1969:

> Scientists attending the 12th annual Pugwash Conference have heard that Argentina appears to be developing the ability to produce nuclear weapons. . . . The mood of the meeting at this Black Sea resort is somber. Some of the more candid speakers from East as well as West have conceded that the arms race has a "life of its own" by being tied toward ever more complex and costly systems through its own momentum rather than by independent political decisions. . . .

We accept the worst in us (war, violence) as inevitable and the best in us as, finally, impractical. And the main thing we are willing to pass on to succeeding generations is woe. Why should they have it any better than we do?

"I agree with you," said one practical man to me recently. "I seriously think that we are threatening the very existence of life. But I'll be dead before anything happens." He was embarrassed by what he had said, but he shrugged.

I am too vividly aware of my own myopia to feel superior. But this summer is to be for me a thrust, at least, in the opposite direction, an exploration into the infinite and an exploration into our species' relation to the planet, to posterity, and to itself.

II

The island has a rich history, and I will start with that. Here, too, as with our just-arriving terns, one can only make notes and look for signs, because the record has its great blanks and its ambiguities.

Great Gull is part of that chain of islands that runs northeastward from Orient Point on Long Island's north fork toward the New England shore—Plum, Great Gull, Little Gull, and Fisher's—reefs of loam and sand and gravel and boulders, all extensions of the so-called Harbor Hill Moraine. These island-reefs may well have been one piece of dry land originally, after they were laid down by the ice of the Pleistocene; now they are fragments tossed on a map of the Sound. On the Coast and Geodetic Survey's navigation charts Great Gull looks like a twisted pickle pointing east and southwest—and little more than a dot of a pickle at that, being only half a mile long, a tenth of a mile wide at its broadest, some seventeen acres in all. Connecticut is seven miles away to the north; Orient Point is seven miles away to the southwest, with Plum Island in between, almost three miles off across tidal rips. The only nearby piece of land is Little Gull, one-twentieth the size of Great Gull; it lies half a mile to the east across another rip.

Before the Europeans arrived, one of the Indian tribes on Long Island—either the Cutchogues or the Manhassets, who seem to have divided between them the outlying islands of the Sound—may have made a summer camp here and feasted on tern eggs and mussels and fish. Possibly (just guessing) it was later used as a fishing station by the Dutch or the English. However that may be, the first human use we know Gull Island was put to involved farming.

In 1675, one Samuel Wyllys received a Patent of Confirmation from Sir Edmund Andros, the governor of New York, for "Plumme Island, together with a small Island adjacent, called Gull Island, the which he hath for many years been in quiet possession of, without Interruption." The islands were granted "as an entire and enfranchized mannor and place of itself," in consideration for which Wyllys was to pay to the crown or to the Duke of York and his heirs or to the duly appointed governor a quitrent of one fat lamb, due each June 24, Midsummer's Day. Wyllys came from a wealthy family of English Puritans who had emigrated from Fenny Compton in 1638, when Samuel was six years old. His father was a leader in Connecticut government, was even briefly governor of the colony, and by the time Samuel was twenty-two (and newly graduated from Harvard College) he, too, was a major figure in colonial government and had married the daughter of another Connecticut governor. He would be a colonial magistrate for more than thirty years, would be elected moderator of the Connecticut General Court a number of times, and would also be chosen a Connecticut commissioner for the United Colonies. In front of the Wyllys mansion in Hartford stood the famous Charter Oak, where a copy of the Connecticut charter was hidden against the possible theft by the same Governor Andros, whom James II sent in 1687 to demand the surrender of the charter. When Wyllys sold the islands in 1688, to Joseph Dudley of Massachusetts (a future governor of that colony), Dudley got the manor and everything thereon, which included, according to the deed, mills, cottages, barns, other buildings, gardens, orchards, pastures, horses, hogs, cattle, and sheep. The price was two hundred and fifty New England pounds, and Dudley assumed responsibility for the quitrent of one fat lamb.

It seems unlikely that much actual farming was done here by the managers that Wyllys and Dudley must have hired; in fact, one wonders just what Wyllys saw in Gull Island in the first place. Plum Island is a different matter. That's a big piece of land, and it is almost within hailing distance of Long Island. Was Great Gull a way station between Connecticut and Plum Island? Or was its original taking just another sign of the land madness that infected white America? But on the other hand, Great Gull may indeed have been farmed then, judging by its later history.

Dudley in his turn sold the manor to Joseph Beebee of Plymouth in about 1700. During the next sixty years the land passed from hand to hand and was split up; the trail of deeds to this place at last breaks off, and when it reappears again, in 1803, one Benjamin Jerom of New London, Connecticut, is selling Great Gull and Little Gull to the federal government. Jerom was said to be an important citizen. He owned a farm on the Thames River in New London, and when he was well along in years he bought the Great and Little Gull Islands as another farm; every spring he loaded a yoke of oxen into a scow and towed them out to the islands behind a sailboat, and then he daily *rowed* the nine miles out and nine miles back to do his farming, or so the story goes.

Perhaps by the time he was sixty, in 1803, that had become a bit more of an effort than he cared to make. At that point, the young national government had assumed responsibility for the safety of coastal shipping and was looking around for places to erect lighthouses. Long Island Sound can be a wicked piece of water, and Little Gull marked the boundary between Long Island Sound and Block Island Sound, to the east. "I know of no situation in this Quarter," wrote one local authority—and he was seconded by a long list of New London shipmasters—"that a Light House can be built on, that will be so useful to the Foreign and Coasting trade, as one on the little Gull Island would be." But Little Gull was isolated. The federal government was more impressed by its isolation than Benjamin Jerom had apparently been—but then, Jerom could go home at night in good weather, and only came out here in the warm months. There would be a good deal of difficulty and expense keeping such a lighthouse provisioned if all the food had to come from the mainland; and Little Gull was barely big enough for the lighthouse and the keeper's dwelling, so Great Gull Island would be needed as a "plantation" for the keeper. Jerom sold the two islands to the United States for $800, and shortly the state of New York put its seal of approval on the project, ceding the Gulls to the federal government. For years, their history would be closely linked.

Abishai Woodward, contractor, to Henry Packard Dering, superintendent of lighthouses for the Sag Harbor district, October 15, 1804:

> . . . [T]he building is carryed up to the heighth of 15 feet from the surface of the Earth, a Well is sunk & stoon'd 21 feet in Depth nigh the Centre of the Island, with a Very plentifull Supply of Water Discharging one barrel in 3½ minutes, and the Water tollerable good, the Sellar dug & stoon'd, the Dwelling House Rais'd and covered with bords & Shingles, I have got on Stone and other materials nearly Enough to Complete the Work. . . . but with Regret I have to inform you,

that the season has been, and still continues so uncommonly boisterous that it is with Difficulty that the Workmen can stand on the Work and they are uneasy and Determined to quit the Work. . . .

The work was stopped for the winter, and the lighthouse on Little Gull was finished and lit for the first time the following year, 1805. The tower was octagonal in shape and was built of smooth, hammered stone, laid in courses like a stone wall. It rose fifty-three feet from its foundations on the dune, which put its light sixty-six feet above the surface of the Sound. Its lantern was a window-enclosed gallery containing a set of sperm-oil lamps and parabolic reflectors imported from Europe, and the keeper trudged up a spiral staircase into the lantern to check them several times a night.

The first keeper was Israel Rogers, who was recommended for the appointment by Superintendent Henry Packard Dering as a good boatman and pilot who knew these waters well and as "much the most smart, active & suitable person" of the three candidates for the job. Of the other two possible keepers, one was a ne'er-do-well, Dering thought, recommended to him by the selectmen of the Town of Groton, Connecticut, in hopes of getting the fellow off town relief; the second was a farmer, not a sailor.

With his recommendation, Dering unwittingly started a twenty-one-year family business in the keeping of Little Gull Light. At least I *think* it was a family business. The record is not absolutely clear. There was an Israel Rogers living in the New London area during that period, and he was married to Serviah Minor Rogers. They had a son, John, and a daughter, Mary, and Mary was married to Giles Holt. The government records show that Henry Dering's Israel Rogers was succeeded by Giles Holt, who was succeeded by John Rogers II. The trouble is, the two sets of records that produce this information don't cross-check. None of the correspondence from the lighthouse district to Washington, including recommendations for successors, mentions the family relationship. (Perhaps this was because it was thought that early Presidents of a brand-new democracy, who appointed lighthouse keepers on the recommendation of the Secretary of the Treasury, would not countenance such feudal passings-on of the job.) Furthermore, the published genealogical records of New London don't mention lighthouse keeping as an employment of either Israel or John Rogers, possibly because neither of them died at that post. There is one neat bit of corroborating evidence, which we'll come to in its proper place in the story, but even that is a trifle flawed.

In any case, having recommended Israel Rogers, Henry Dering also urged the Secretary of the Treasury that the keeper

> have the benefit of the Great Gull as was originally intended, [since] this island can be of little or no service to any other person and of little value to the United States. . . .
>
> It will however be of essential service to the keeper for a garden, pasturage for a cow or two, to raise a little Indian corn upon by manuring and cultivation. . . .

He suggested that the keeper might also cut hay on the island; that way, he would have fodder for his cow in winter, when he would have to bring the animal over to Little Gull, "as the Great Gull is so difficult of access during the winter season that it cannot be visited unless at particular times."

This was a bleak station, especially after the summer was over. In December, 1805, Dering visited the lighthouse, and the keeper's wife told Dering "she had not seen the face of any human being excepting her own family for more than two months. . . ." Keeper Rogers was already threatening to leave after his first year's service was up, particularly if his salary weren't raised. He was then earning $333 a year, tops for a lighthouse keeper but low, nonetheless. He told Dering "he should try to 'weather it' (to use his own phrase) for one year" but thought he would quit after that, because his and his family's sacrifice was too great. ". . . [H]e had rather keep the light house at Eaton's Neck [on the Long Island shore] for 200 Dolls than the Gull Island for 400." Secretary of the Treasury Albert Gallatin proposed to President Jefferson that Rogers's salary—and that of two other federal lighthouse keepers—be raised to $433 a year, and Rogers decided to stay on.

The following winter and spring, Rogers reported, severe storms and high tides did "great dammage here."

> The bank is washed away quite up to the celler door and the Island is much broken in different places & washed off—
> It will be with the utmost difficulty that any Oil can now be landed and as the land has been so much washed & torn, I fear exceedingly that unless something is soon done to repair & prevent further breaches, it may by & by be too late.

Dering was ill at the time, but he sent an agent, Nicoll Fosdick, to inspect the damage. "When we Surveyed the Island and fixed on that spot for a Light House," Fosdick reported with surprise, "we expected in time it would be necessary to have the Banks secured by a wall, But had no Idea it would be in our Day, and the fact is, that more of the Island washed away last winter than has for Twenty or Thirty years before. . . ." Fosdick believed he knew one reason: during the construction of the lighthouse and the keeper's dwelling, all the vegetation on the island had been trampled and killed, which exposed the soil. Fosdick suggested that no "pig or fowl run at large on the Island untill it is fairly Sorded over, and Some Trees or Shrubs must be made to grow on the Sides of the Banks. . . ." Meanwhile, the immediate problem was serious; the foundations of the dwelling house were in danger of being undermined in the next severe storm, and so was the vault for the sperm oil. Fosdick supervised the building of a seawall to prevent this, and when it was done remarked that "I flatter myself, it is now secure, I think the wall will Stand as long as time. . . ."

Israel Rogers kept the light until November, 1809. His successor was his assistant, Giles Holt. Holt was about thirty years old when he became keeper; he had been married to Mary Rogers for five years. I've seen a miniature portrait of him. He was a spirited-looking young man with blue eyes and dark brown hair, curly in front and pulled into a pigtail behind; he wears a gold earring in the portrait, and a high white collar, white tie, a vest, and a blue coat—reproduced in the portrait by a cutout of silk pasted to the painting. He did some farming here on Great Gull, because in July, 1810, Dering worriedly wrote the Secretary of the Treasury that "by frequently plowing and tilling the Great Gull Island . . . it breaks the sward & turfs up & occasions the Earth to waste and blow away so much that in the course of some years it will probably become a mear sand spit or reef of rocks." Another reef was not wanted in Long Island Sound, Dering thought, so he had ordered Holt not to do any more plowing on Great Gull except in a small garden and a corn patch. He "proposed to the Keeper in lieu of the profit that might arise to him from tilling the ground to keep a small flock of sheep . . . which can remain there during the whole winter without fodder. . . ." Holt agreed, if the government would provide him rails for fencing the pasture, and Secretary Gallatin endorsed that purchase.

It was singularly important to keep lighthouse keepers happy when they manned such lonely stations as Little Gull. In 1811 Superintendent Dering reported to Gallatin that the Holts, "who are very decent people," had "petitioned me very hard to have a small addition of two small bedrooms of only 10 by 12 feet" added to the ground floor of the keeper's house. At the time, the house consisted of two little rooms on the ground floor and a single room in the half-story above. That might have been fine for an unmarried keeper and an unmarried assistant, but such wasn't the present case; the Holts and their children and the assistant keeper, *his* wife, and *his* children were all crammed together in the three rooms, and they were decidedly unhappy. Holt threatened to leave if something weren't done; he "pressed me so hard . . . & it appeared to me so necessary that I gave him encouragement that it should be built." Gallatin agreed, and in the spring of 1813 the job was finished.

In a sense, the demand was met in the nick of time, for if the

Holts had been dissatisfied with their accommodations in the summer of 1813, they would probably have abandoned the station as soon as the War of 1812 spread to Long Island Sound. The British had been blockading the Chesapeake and Delaware bays, and early in the spring of 1813 the blockade was enlarged to include the approaches to several other centers of commerce, including New York. A squadron under the command of Thomas Hardy, one of the heroes of Trafalgar, took up a station at the eastern entrance to the Sound. Commodore Hardy was evidently hard put to know what to do about Little Gull Light; it was, after all, as useful to him as it was to American shipping. But an American squadron under Stephen Decatur made a run for the open sea out of New York and couldn't get by; Decatur put in at New London harbor and then sent a few boats across the Sound to Long Island. Officially the Sound itself was not yet included in the blockade, but under the circumstances Hardy apparently felt he'd best do what he could to hamstring any possible American maneuvers there. So he sent word to Holt to extinguish the light. Holt bravely replied that he'd keep the light lit until his government instructed him otherwise. So the commodore sent a party to the island to dismantle the lantern; they took away the thirteen lamps then in use and destroyed an older set that had been stored in the light. The keeper's house was searched, in case the Holts had a large enough supply of candles to substitute them for the lamps, and when the sailors left they warned Holt that if he managed to set the light again, they would come back, break up the lantern completely, and wreck the oil cistern.

The keeper hurried off to district headquarters, at Sag Harbor on Long Island, to tell Henry Dering what had occurred. Dering sat down immediately and wrote the Secretary of the Treasury. The superintendent thought the British action equivalent to "putting out both of their eyes for one of ours—as their barges frequently get lost. . . ." He believed they would want the light lit again before long. Whether they did or not, Holt was still on Little Gull in January, when the British commandeered a few gallons of his oil, which they paid for in Spanish coin. Holt was careful—in fact, *anxious*—to turn this money over to Dering; he made such a fuss about it, I wonder if he didn't have a special reason for wishing to prove his honesty and faithfulness. According to a story current in New London more than a hundred years later, one Serviah Rogers—who likely was none other than Holt's mother-in-law—had been selling supplies to the British squadron, and very profitably, too. That was not considered much of a crime in New England during the war, but it would not have sat too well with the federal government.

Next month, the British were back, and once more Holt sailed off to Sag Harbor. ". . . I am sorry to state, [he] informs me the enemy have made him another visit Tuesday last," wrote Dering to his superior, "& told him to leave the island for it was their determination to distroy all the public property belonging to the establishment." They had already taken away a lot of miscellaneous equipment, Holt said, along with several hundred gallons of whale oil and twenty-two bushels of coal—presumably the fuel that was burned in a stove in the lantern during the cold months to keep the oil warm enough to flow into the lamps. And they had also robbed Holt and his assistant of a good deal of personal property, including firewood, lumber, tools, and all of the "fish carrs" that Holt used for storing his catch alive and fresh. (Selling fish would have been one way a keeper eked out his living.) Holt admitted he was now concerned for his own safety, but Dering asked him to stay put, "that if he left it, they would certainly burn the House and destroy the property. [He] promised to continue until taken off or the house destroyed." But if Holt did stay, it did no good. The British—"the Goths," Dering once called them as he retold the story—came onto Little Gull and wrecked the place. They broke all the glass in the lantern, tore off doors and all the rest of the lantern equipment and carried everything usable away, including door locks and the brass petcocks on the cisterns; in the keeper's house they broke windows, knocked down partitions, carried away doors and closet shelves and other boards, and stole the windlass, bucket, chain, and rope out of the well. Judging from Dering's description, they went at the buildings with sledgehammers.

Five hundred rails were taken from Great Gull; whether these came from a sheep corral or were still just stacked, I don't know, but Giles Holt was probably not doing any farming at all on the island by then. The half-mile of rips he had to cross often kept him away in bad weather, and that discouraged him. Furthermore, if he had been pasturing sheep here, the British would certainly have relieved him of the animals, and he would have said something about it to Dering.

Dering's son came ashore on Great Gull after the war, while repairs were being made to the Light House Establishment, and he found two wooden grave markers. He jotted the inscriptions in his journal:

SACRED TO THE MEMORY
of
RICHARD THOMAS
late a private Band Marine
H.M.S. Superb
killed in a boat
Aug. 4. A.D. 1814
Aged 25

View stranger, where this poplar grows
A Britain lies beneath the sod,
Who fought and died in England's cause
A debt he owed his king and God.

And

SACRED TO THE MEMORY
of
MR. JOHN ENGLISH
late
Boatswain
H.M.S. Superb
Departed this life Aug. 4.
A.D. 1814

August 4. The peace talks at Ghent would be seriously un-
derway in only a few days. Killed in a boat; "fought and died in
England's cause"—I wonder what the mishap was. Were they
shot at—or were they run down in a fog, their dinghy crushed
by the *Superb*? They were a long way from home.

On September 23, 1815, a hurricane hit the southern New
England coast and plowed northeastward. No one, of course,
was prepared for it. The wind-driven water rose so high it stood
eight feet deep in the streets of New London. The wind tore off
roofs and blew down barns at Orient, on Long Island, and the
high-water mark there was well inshore of the previous record.
At Providence, where the worst damage was done, the rising
water and the wind wrenched ships from their moorings, drove
them into a bridge and destroyed it, swept them into town
where they knocked down buildings. In New Bedford Harbor
every ship but two was hurled up on shore. Sixty ships in Boston Harbor were wrecked or damaged, and salt spray blew far
inland. It was the worst storm anyone had ever seen in New England, and it remained graved on the memories of those who experienced it. "I have," wrote Oliver Wendell Holmes, "besides
more specific recollections, a general remaining impression of a
mighty howling, roaring, banging, and crashing. . . ." He re-
membered people trying to get down to the banks of the Charles
River in Boston, "but they were frequently driven back, and
had to screen themselves behind fences and trees, or 'tack'
against the mighty blast, which drove them like a powerful current of water." If it was bad on land, it must have been terrible
at sea. At Little Gull, the waves crashed completely over the
island and scoured its margins until there was alarmingly little
of it left. At one point, according to an old story, a pig was
swept out of its sty near the keeper's house and was saved from
drowning only because it happened to be washed within an arm's
length of a window in the house and was hauled inside.
From what he called the "Remains of Gull Island," two days

after the storm, Giles Holt sent his superintendent a letter, courtesy of a passing vessel, and described what had happened at Little Gull. The seawall, which Fosdick had confidently predicted would "Stand as long as time," had of course vanished, been swallowed up by the sea. The oil vault had been swept away; so had the foundations under one end of the dwelling house. Half the "upland" on the island was gone, leaving the light and the keeper's house perched on a small, elevated patch above perpendicular dunes as much as twenty feet high. There were only "3 places that a man [can] get up or down & that with much difficulty." The lantern, just freshly repaired, was again in ruins—only one pane of glass left in the windows, all the reflectors behind the lamps down and some of them whipped away by the gale. Holt recommended that everything valuable be immediately removed from Little Gull, and he warned Dering that "Should their come what we generally [call] a common storm it will be impossible to [remain] here."

However, Montauk Point Light was also a shambles, and Dering had already gone there in response to an urgent message from the keeper, so he didn't get back to Sag Harbor until three days after Holt's letter arrived—

> in which time [Mr.] Holt became so anxious for his personal safety & that of his family, He left the Island & came to this place & informed me he would not re[main] on the Island for 1000$ pr. Annum in its present situation. That himself & wife had not slept since the storm from fear.—That he should leave it & the remains of the public property would be much exposed after he was gone.—I therefore hired a Vessel to go on & remove him & family and all the public property off.

At the urging of Dering and others, the federal government agreed to build a massive seawall on Little Gull; but the money would have to be appropriated and bids asked for, and with winter approaching there would be no way to get started on the project for many months. In October and November three ships went aground because the light was out, while Dering was having temporary repairs made and the light put in working order again—with the idea that Holt would return to his station. Holt's patience was worn thin; he had put up with a great deal in the past two years. He petitioned President Madison for an increase in salary, in a document that seems to have been prepared for him by someone trained in the art:

From a sense of duty I owe to myself and family, I cannot think of resuming my trust as Keeper without stating to the President the great privations I shall be obliged to suffer, in addition to those myself and family have ever experienced since we have resided upon the island.

. . . [A] person now residing there, must be exposed to a more than common risque of his life, in the event of another such [high] tide.

Your petitioner cannot think of taking on to the Island with him his Wife and family untill some more effectual work of defence is made for the security of the Island—this Circumstance will greatly add to my former privations . . . , which have at all times vastly exceeded those of any other Light House establishment in the United States according to the best of my information.

I have never had, and must now never expect to have the comforts of a Garden, not the raising of a potatoe or any other kind of Vegetable, or the privilege of keeping a Cow, which adds greatly to the support of a family of Children.

In addition to this every article for the support of myself and family have greatly increased in Value since my appointment of Keeper of the establishment.

The difficulty and expence of procuring and getting on to the Island every necessary article for the support of a family, and that of fuel is very great, and many times attended with much cost, labour and risque.

The privation myself and family suffer on account of Water perhaps is one of the greatest sacrifices, having no other than rain water that is fit for use, except imported.

I am also Obliged at all times to keep and pay, from my

own Salary an assistant Keeper, and find it very [pressing] under all these difficulties to support myself and family.

The adjacent Island is so distant and difficult of access as to render it almost intirely useless for any imp[loy]ment, and the situation of the establishment is such that it affords me no local advantages of profit and I do not derive in my present situation any other emoluments [than] what results from my salary. . . .

His salary was raised by a hundred dollars, and Holt returned to keep the light. But he would not stay long. By the following summer, as bids on the improvements at Little Gull were let and considered, Mary Holt had evidently rejoined her husband, and she could no longer stand the place. "Mrs. Holt has becum so uneasy and alarmed in her present situation hear," Holt wrote to Henry Dering, ". . . that I find for her peace & comfort I shall be compeled to resign my appointment. . . ." He said he did this reluctantly, particularly since the improvements the government intended would make the place much safer than it had ever been. "But the [te]rrors of Mrs. Holt's Mind are such as sensibly to affect her health their fore for her comfort and in hopes [it] may restore her to better health I am induced to [give] up the duties of my appointment and quit [the] Island."

So in late September, 1816, John Rogers II took Holt's place. The stout seawall was built; eight to ten feet across, buried ten feet deep in the island, and rising to the level of the remaining upland, it circled the lighthouse and the keeper's dwelling. The outer course of the wall was made of square blocks fastened together top to bottom with copper bolts, and wherever it rested on ledge, the wall was bolted to that, too. When the foundations were dug, the earth was thrown outside the circumference of the wall—putting back some of what the sea had taken away—and after the wall was finished Dering strongly urged his superiors in Washington that this earth be ballasted with stone brought from Great Gull to protect the foundations. That was done several years later. From time to time this ballast would have to be replaced, but the island was at last relatively secure. A barn was raised on Great Gull. The keeper's house was enlarged and improved and given a new foundation. When in 1821 another September gale struck the Sound—a tempest sufficient to make the lighthouse tremble, John Rogers said—the Light House Establishment came through unscathed, except for the barn here on Great Gull, which was blown down.

John Rogers seems to have had a far quieter tenure at the light than Holt did. The official records of the period mention Little Gull only occasionally. The best story about his stay, in fact, surfaced in *The Day*, New London's daily newspaper, in 1920. Serviah Rogers, former trader with the blockading British, was running a tavern near the waterfront in New London; it was a rough-and-tumble sort of place, but she was in charge, with her husband taking his ease in a back room. Her customers called her Squire Viah. In that, if one listens close, one can hear not only high praise—backhanded Yankee style—but also the well-known Yankee accent, which would pronounce it "Squiah Viah." Squiah Viah was a tough old bird; no doubt about that, and local sailors were so much in awe of her that her reputation long outlived her. In the story told in 1920, her relationship to John was that of a sister, not a mother; but the story was a century old by then. Any blood brother of hers would have had to carry her maiden name, which was likely not Rogers; thus John Rogers was doubtless her son. Nonetheless the story went that Squiah Viah

> had a brother, John, who kept the light at Little Gull Island and to whom she was so strongly attached that she would take a sail across the Sound to spend the day with him and break the monotony of his solitude. Sailors told of her remarkable skill in handling a sailboat and of her prodigious strength at the oars when the wind died out. One sailor knew for a fact that the squire made evening trips to Little Gull and that she

walked on the water going and coming. One smackman told of coming down the Sound one wild night with topsails reefed and the wind blowing a gale from the southwest. Through the haze he saw a human form crossing his bow and as his craft drew nearer he saw that it was Squire Viah with a lantern in one hand and a staff in the other. He asked her if she wanted a line flung out and the reply was to the effect that she wouldn't bother. She was going over to the island to call on her brother John, and she guessed she was good for the journey.

Frederick Chase became keeper at Little Gull after John Rogers was fired in 1826. There had been a lot of talk about the poor light that Rogers kept, which wasn't entirely his fault, because the lamps and reflectors at Little Gull were of very poor quality. But by the winter of 1825–1826, he had fallen into the habit of leaving the island pretty frequently. He had trouble keeping assistants, so whoever he left in charge was not likely to have much experience, even if that assistant took his responsibilities seriously. Rogers was warned to mend his ways, but in the dark hours of Christmas Day, 1825, he was away again, and a vessel out of Sag Harbor nearly ran aground on Little Gull because the light was out. The master of the ship entered a formal complaint, and Rogers was discharged. In March, Fred Chase began his thriving tenure.

Chase's great-grandson, Clarence Ashton Wood, wrote a piece about him for the *Long Island Forum* during World War II. Wood said that the keeper was related to Salmon P. Chase, chief justice of the U.S. Supreme Court, 1864–1873. A Seventh-Day Adventist, he was evidently one of those devout social reformers with which New York state was so rich in his day; he owned property on Shelter Island, near Greenport on Long Island, and he had a dream, never fulfilled, of establishing there a community, to be called Sobrie. He was a respected citizen who was addressed as Squire for more conventional reasons than Serviah Rogers was: he was a justice of the peace, school trustee, local legislator on Shelter Island; he was evidently moderately well educated, wrote some Latin, knew at least a little Greek, and seems to have collected aphorisms in various languages.

He was the only keeper at Little Gull who is known to have kept a journal, and much of that journal has survived. His entries always began with the Latin name for the day—Dies (or D.) Solis for Sunday, Dies Lunae, Dies Martis, Dies Mercurii, Dies Jovis, Dies Veneris, Dies Sabbatii. He made marginal notes to himself sometimes, in a middling sort of Latin. "Primo Larus Ova"—the first "gull's" eggs of the year—was one such. "Primo Assellus"—the first codfish—was another.

Sometimes the marginal note is difficult to translate. "Alteno animo sum ab illo (VIZ.) Edvardus," he commented once. I didn't get past second-year Latin in school, so I am handicapped; Latin scholars tell me the keeper's cases are a little muddled here, so I am on my own. My first problem is that there is no such word as "alteno" in Latin. Squire Chase could have meant "alterno" or "alieno"; the translation might read either, "I am encouraged to change my mind about Edward," or "I do not agree with Edward." In either case, there was some problem with Edward Conkling, his hired man at the time. Several months later he used the term "Edvardus Ludibundus"— Edward is full of play. Was Edward in trouble with his boss again? Apparently not. A very uncharacteristic series of marginal notations and journal entries begins at that point, September 5, 1827, and runs for eight of the next ten days. "Ludibundus," says the margin. Full of play. "Imployed in Nothing," says the journal, or "imployed in work & play." Never in the next five years does the journal reflect anything so spiritedly larky in Chase's nature. He seems to have been hard work and Sobrie itself, the sort of man who keeps exact accounts of things: the lengths of time, to the day, his many assistants stayed with him at Little Gull Light; the numbers of lobsters caught, the numbers of blackfish and shad and codfish and mackerel and bluefish; the numbers of tern eggs gathered, the bushels of potatoes dug, and ears of corn shucked. His great-grandson, writing the piece for the *Forum*, had beside him Chase's journal for 1833, which I have not seen, and he reported that the keeper's then hired man, Eli Manwaring, had to bullyrag Chase into agreeing to cut up a wooden buoy they had found on the shore of one of the Gulls a year earlier; they badly needed the firewood, but Squire Chase thought the owners might still want their buoy, and he was loath to destroy it. He compromised by keeping the iron fittings of the buoy "for the owners when called for."

It is nice to think of what Great Gull was like before the fort was built, nice to think of the relationship of Fred Chase to the soil of the island before it was piled and cemented and filled with rubble and edged with thousands of tons of granite. I try to see his Great Gull through the ink faded to brown on the brown pages of his tattered journal. Judging from a navigation chart of the 1850s, and from the profile of the island included on the chart, it had beaches, dunes, a prominent reef at the end nearest the lighthouse—the reef even more lethal-looking than it is now. The island had fields for corn, potatoes, barley, hay. Either here or on Little Gull (Chase's journal is a blank as to that, but it was almost certainly here) he had a garden where he grew such items as cabbages, onions, parsnips, broom corn, beets, peppers, squashes, tomatoes, beans, cucumbers, saffron, sage, summer savory, sweet potatoes, melons, and what he called Buta Baga. It was by no means ideal farmland: "very tough ploughing," he noted on May 9, 1827, and that July 2 he added that he had just spent the day "Imployed in Hoeing the foulest piece of Land that I ever saw without any exception." A small orchard stood somewhere on the island. He already had apple and pear trees here when in April, 1830, he planted cherry trees and currant bushes on his Great Gull Island plantation, along with cedars and willows and buttonwoods. There was rail-fenced pasture, and he kept cows here, and oxen, and a flock of domestic geese, and possibly his few pigs. Great Gull had a well by then—too near the beach, because severe storms would sometimes fill it with salt water. The barn that had blown down in a gale during John Rogers's term as keeper had been restored; Squire Chase reported its being blown down again in another gale. He built a house here. "Began to fix the timber for the House to be put on the other Gull," he wrote on February 28, 1830. On August 11 he wrote in big letters that he had "Rais'd the House on the Large Gull Island." It was tight enough by December, for when a northeasterly gale drove a sloop on shore here, the master of the sloop and his passengers evidently made it their home for eight days, until they could float their ship again. The next March, Chase "carted stone to pave the house on the other Island"; that August, having "Finish'd Paveing the floor," he "began to point with lime mortar the wall of the house." "Finish'd pointing the house inside and out to day," he reported on August 11, 1831, "It being one year from the time it was rais'd." That was the sort of timing he enjoyed. The next day he painted the house, and on September 22 he and a friend or relative "stay'd all night in the New House for the first time since it was built—gather'd corn &c."

While he and his hired assistant and perhaps a son or some other partner did their building, their plowing and planting, and while they hoed and weeded and harvested, they also netted and lined for fish and hauled lobsters and collected thousands of tern eggs—all while seeing to the running of Little Gull Light. It is hardly any wonder, then, that when his hired man had to go to Long Island to testify in a court case one July, Chase should complain that this "puts my work behind very much. . . ."

The Gulls were still very still very isolated, though a shade less so than before; commerce was slowly building on the Sound, and the early Sound steamers were now covering routes between New York and New England. But almost all ships were still driven by wind; certainly any craft Chase would have had used sails, or oars. Depending on the weather for propulsion meant waiting for a fair wind, and that fact alone isolated the lighthouse. On April 5, 1830, Chase was at Shelter Island with a load of firewood for Little Gull Light, and there he lay for thirteen days while the wind blew out of the northeast—blew in his teeth, in other words, since he had to sail northeast to reach Little Gull. There was no sense beating back and forth, zigzagging all the way to Little Gull if he didn't absolutely have to. Not in the "Devil's Belt," as sailors once called the Sound. So he waited. On the seventeenth the wind went around to the south, which would have let him fetch the lighthouse on one long reach if it held; but it didn't. After he had started out, it shifted back into the east-northeast and blew hard, so he made for the nearest harbor. Not until the nineteenth did the wind get into the west, and he had a gentle following breeze to push him down eastward. "Arrived," he wrote in big letters, "at the Little Gull at 4 P.M. after having waited 2 weeks for a good time to come down. . . ."

Almost all his supplies had to reach him over water: the fuel and wicks and reflectors for his lamps, most of the firewood (although he did collect driftwood from the shore), his flour, most or all his butter, his lumber, a good deal of his lobster bait, clothing, mail.

But if he was isolated, Frederick Chase surely was not lonely. His wife Rebecca and many of their nine children and *their* spouses seem to have spent a great deal of time on Little Gull; if the hired-man-cum-assistant-keeper happened to be married, then *his* family might be there, too. In addition, the lighthouse seems to have served as a frequent stopping place for passing sailors seeking fresh fish and lobsters, a bed for the night, or a chat. And there is much reason to believe that Chase ran a kind of boardinghouse on Little Gull; he had numerous guests whom he had never met before they arrived, and taking in boarders, as well as selling fish and eggs and vegetables, was how lighthouse keepers padded out their incomes. I suspect that this was one reason he built the house on Great Gull.

In winter they shot sea ducks; Chase repaired his lobster traps or made new ones, collected driftwood for his stoves, dug potatoes on Great Gull, butchered his hogs. In early spring he set out his lobster traps and one or two fishnets, started plowing with his yoke of oxen, and so began the busy months. In the fall, he cut and stacked his cornstalks, probably stored the last cutting of hay in the barn on Great Gull, took up his traps and nets, and settled in for the winter.

All the while, he fussed over the lamps in the "Lanthorn," as he called it. There were fourteen of these lamps, each with a wick inside an arrangement of glass tubes that insured a draft. The whale oil was now supplied in two grades, summer and winter, because cold weather thickened the regular oil too much; even so, the lighthouse keeper still had to keep a stove fire going during the winter to keep the oil warm and flowing. Chase evidently checked his lamps more than once during the night, and sometimes frequently, when they were giving him trouble. They often did.

> Dies Martis 2d [January, 1827] Fresh winds from the N.W. to W. attended with snow squalls and cold weather the Lamps did not burn well although there were three fires made During the night. . . .
>
> Dies Mercurii 3d . . . Made 5 fires in the Lanthorn *viz.* at 4-9-12-1 and 3 O'Clock in the Morning — The Lamps burnt Middling well this morning but not so well as in December when there was but 2 fires made in the Course of one night. . . .
>
> Dies Solis 22d [April] . . . the Lamps burnt well the night before Last but Last night they did not burn so well Occasioned by a fowl flying through the Lanthorn (Viz) a Quack.

The Quack was probably a black-crowned night heron or a black duck. Birds, then as now, were attracted to the light in the lantern at night, and from time to time they hit the glass. On January 11, 1828, he reported:

> Last night at 2. O'Clk in the morning a flock of Sea Ducks came in contact with the East Square of the Lanthorn and broke in 4 panels of the Glass to atoms while we were going into the Lanthorn and about half way up to the Lamps — Which made a most tremendous Clashing with the glass &c. 1 Duck came in to the Lanthorn being a species called Redheads. . . .

He had a low opinion of his lamps. The journal is sprinkled with such comments as "the lamps burn well as they usually do with poor Oil," "The Lamps burn well—for them————or for such Lamps," "The lamps burn as well as usual— But without an Hyperbola the man who approved of the Apparatus which is now in use at this Establishment must be M a d ————"

Chase wasn't the only one to complain about the apparatus in use in lighthouses. Henry Packard Dering's son Henry Thomas Dering was now the superintendent of the lighthouse district, and ever since the early 1820s he had been suggesting to the powers-that-were in Washington that something ought to be done for the light on Little Gull. But in general United States lighthouses lagged far behind their European models in terms of equipment, and they did so for generations. When Frederick Chase was a keeper, American beacons used a system of lamps and reflectors "invented" in 1812 by a lighthouse keeper named Winslow Lewis, after Lewis had visited several European lighthouses and seen the lamps they were using. Aimé Argand, a Swiss, had developed the system of burning a circular wick inside a glass tube, which brightened the fire and cut down the smoke. He also placed parabolic reflectors behind the lamps, which intensified the light. What Winslow Lewis did was to steal those ideas, reproduce them badly, and apparently try to combine them with a crude sort of lens. Lewis' lenses were worse than useless, and if they were ever part of his working apparatus they were soon got rid of. But what was left was no great shakes. Arnold Burges Johnson, who was chief clerk of the United States Light-House Board in 1889, described the reflector of the Lewis light as follows:

> [It] was . . . a thin sheet of copper, commonly segments of a sphere, plated over with a slight film of silver, though the copper was so thin that the compression between the arms of its iron supports materially altered its form, and its silvered concave surface had much the grain and lustre of tinware, and would reflect no distinct image. The patentee of 1812 made no pretension to a knowledge of optics . . . and his reflectors came about as near to a true paraboloid as did a barber's basin.

That was the sort of thing Chase contended with, and his lamps seem to have been rather worse than most. He nursed them, boiled them out every now and again, polished the reflectors, trimmed the wicks or put in new ones, replaced the tubes, which frequently broke, and probably tinkered with the focus of the reflectors, because to get the brightest possible light out of the apparatus a keeper had to screw the lamps forward, away from the heat-sensitive reflectors, in warm weather and back in cold weather.

The job of supplying the lighthouses was let out to Winslow Lewis and, through him, to various subcontractors, and once a year a boat would arrive carrying Little Gull Light's supply of oil, which was stored in two cisterns on the island; the shipment would also include wicks and tubes and "Buff skins"—presumably a kind of chamois for polishing the reflectors—and yards of cotton cloth for cleaning the tubes and the windows of the lantern.

On at least one occasion, the contractor stayed to tinker with the lamps himself. "The contractor Mr. Swain said that he thought the Light Burnt rather Dull Last night," wrote Chase. "But the Lamps burnt very well when I put them out this morn.

. . ." The next day his injured pride turned to exasperation. "The Lamps burn poor after being fix'd anew 2 being out and 2 about out—1 that had a new burner and barrel put to it yesterday leak'd all the oil out this morning—& 1 the thimble broke in turning it up—which has rendered it useless. . . ." The following day he added that "we have not got the Lamps yet to burn so well as they did before they undertook to fix them. . . ."

From time to time Henry Thomas Dering or one of his subordinates would visit, and these were not uniformly happy occasions. "The Superintendent came here to Day," Chase noted on August 13, 1829. "Found Quite a Deal of fault with the Cleanliness of the Lanthorn." He recorded only one other instance, however, of faultfinding by his superiors, and that stemmed from his custom of leaving someone in charge at the light for a week or more at a time and going to Long Island— sometimes to take care of business, sometimes just for a rest and a bit of eel fishing in Great South Bay. That was the way John Rogers had got himself fired. "Hear'd to day Something about being off too much," he noted in the margin for September 16, 1832. But he was probably as conscientious as any of the under-paid keepers of the 1830s. And he had fault he could find in his turn. He once went six months without a paycheck, making fruitless trip after fruitless trip to Sag Harbor to "git" his money. And despite his numerous complaints to Dering, and Dering's to Washington, the federal government didn't get around to replacing the lamps in his light during his entire ten years as keeper.

Chase's character makes him seem kin to the present tenants of Great Gull Island. He was sufficiently interested in the curiosities of nature to note in his journal the spider spinning its web in the lantern in December, the blooms of red clover and wild turnip and "Chadlock," probably a mustard, that he picked on Great Gull at the end of a December warm spell in 1829, with "Large Grasshopers flying about" as he walked; the albino robin someone brought out to him—"A very Curious bird in

deed it being of the robbin species but all over white as milk";
strange fish, strange birds, strange weather phenomena, the appearance of things around him—"the water smoked this morning like a Pot of boiling water"; the seals that swam near the
islands very occasionally and once even hauled out on the eastern end of Great Gull. I wonder if he didn't even do some tern
farming for the sake of his egg crop: he mentioned burning over
the beach grass on Great Gull one year, a few months before the
terns arrived.

Great Gull and Little Gull changed appreciably in the next
sixty years. The lighthouse was torn down and rebuilt in granite
after the Civil War. The new tower was ninety-one feet tall. At
the same time, a new keeper's dwelling was built; judging from
a drawing made in 1892, it was a three-story house, almost half
as tall as the light tower. (The house has since burned down and
been replaced by a low, characterless structure.) By then the
light had three keepers—a chief and two assistants—and lard oil
was being burned in the lamps. Outbuildings and equipment
proliferated; steam sirens were installed as fog signals, a landing
was built, and a wharf, and a small harbor; by the 1890s the
lamps burned "mineral oil"—kerosene—and the staff of the light
included a third assistant keeper. The farming diminished on
Great Gull, although many keepers, apparently, pastured cows
here. Basil Hicks Dutcher, describing Great Gull in *The Auk* in
1889, remarked that it had been "purchased by the Government
to serve as a garden for the keepers . . . , but it was so overrun
with mice that it was [now?] useless for that purpose." (In fact,
among the hordes of mice on the island, Dutcher had discovered
in 1888 a new species, which was given the name Great Gull
Island Meadow Mouse.) "And now," he went on, "its sole use is
as a breeding place for the Terns and as a convenient and suitable spot for credulous people to search for the buried treasures
of Captain Kidd." Dutcher also said that by 1889 the island was
"covered by a growth of coarse grass, with here and there a

small clump of bushes." Whatever happened to those groves and orchards of the 1830s? Storms? Girdling by mice? "In a hollow on the north side of the island," he noted, "is a small fresh-water swamp, dry and overgrown with cat-tails in the fall." Was the swamp all that was left of the old well?

That year of 1889, with the bird-conservation movement just reaching full flood, the American Museum of Natural History in New York City sponsored its first expedition of any kind, a week's visit to Great Gull Island by William Dutcher (Basil Hicks's father) and Frank M. Chapman. Both Dutcher and Chapman played major roles in advancing bird protection. By 1889, wrote Chapman later, "all that were left of countless numbers of [terns] which once inhabited the shores of Long Island were to be found on Great Gull Island. . . ." And that colony was not doing well, suffering from eggers as well as plume hunters. From a population of three or four thousand birds in 1886, the number dropped to a thousand in 1893. So with the one remaining known colony along the Sound in serious trouble, a consortium of New York City organizations moved in to defend it. There were now state laws to protect terns; all that was needed was enforcement. The Linnaean Society of New York, the American Society for the Prevention of Cruelty to Animals, and the West Side Natural History Society hired the lighthouse keeper at Little Gull Light, Captain Henry P. Field, to watch over Great Gull, and they arranged with the state of New York to make Captain Field a State Game Protector. With that kind of close watch, the colony recovered rapidly. William Dutcher, then chairman of the bird protection committee of the American Ornithologists' Union, estimated that by the summer of 1896 there were seven thousand pairs of terns nesting on the island, and reported with satisfaction:

> That the colony has grown very largely is evidenced by the fact that an overflow colony of some hundreds of birds has established itself on the north end of . . . Gardiners Island.

. . . The keeper of Montauk Point Light informs me that the
terns were more numerous about the point during the past
summer than for many years.

Dutcher's committee proposed to "continue the protection of
this colony until, if possible, the south side of Long Island is
again populated with these beautiful birds, as it was before they
were practically exterminated in 1886 by the demands of fashion."

But the role of the island as a seedbed for tern colonies was
just about over. That highly successful year of 1896 the United
States was becoming a world military power; the War Department wanted Great Gull as a site for a coastal-defense fort, and
as the Secretary of War and the Acting Secretary of the Treasury discussed the transfer of the property, they came to the remarkable conclusion that though the island had originally been
set aside as "a light-house reservation," it had "never been occupied for the purpose for which reserved." In April, 1897, as
the terns were arriving, J. W. Hoffman & Co., contractors, of
Philadelphia, moved onto the island and set up their equipment—a wharf, six derricks, a boiler house, storage and tool
sheds, blacksmithy, oil house, concrete mixer, water tanks,
stone bin, stable for twelve horses, hay shed, ice house, a big
dormitory and dining place, eight workmen's shanties, an office
for the contractors, three shacks for the government engineers, a
small power plant, railroad tracks, and piles of building materials. All this was reported sadly to *The Auk* by one J. Harris
Reed, who visited on Great Gull between the end of August and
the first of October and interviewed the men there. Reed wrote:

> The whole plant took up over one-half of the area of the
> island, leaving only a small portion of the two ends for the ac-
> commodation of the Terns, who were compelled to divide
> themselves into two distinct colonies of about one thousand
> birds each. In these crowded quarters they congregated and
> laid their eggs, some in the grass, while others took to the bare

patches of sand and tops of large boulders along the beach. No
sooner had the workmen discovered [the eggs] than they began
collecting them for eating purposes, as fast as they were laid.
This was principally done by the negroes and Italians, who
provided their own meals, and I was told by them that in
some instances as many as a dozen eggs were eaten daily, by
an individual. A great many were also collected out of curiosity, [and] were blown and carried away as keepsakes. On one
occasion, a New York man visited the island, and collected a
large basketful, which he was permitted to take away with
him, with a promise not to return again. The crews of the vessels which landed there also participated in the shameful
work. . . . I would say that it was almost impossible for
Capt. Henry P. Field, or any one else, to do any protective
work, under the circumstances, . . . for most of the depreda-
tions were done about daybreak, before the officials were up.

Based on what he was told, Reed judged, half the two thousand
birds and possibly more had deserted before the nesting season
was over, and the rest left late in September, following a bad
storm; the colony, he thought, had produced almost no young
birds. He added:

I am also informed by good authority that the Government in-
tends erecting another gun on the east end of the island; if
such be the case, it will consume all the earth from the remaining portions of the island, to form the breastworks,
which will virtually leave nothing of Great Gull Island
beyond the fortifications, and will completely destroy it as a
resort for terns.

The following spring, war was declared with Spain, and people living along the coast began "hearing guns of Spanish
cruisers and seeing ghosts of hostile battleships," in the words of
Colonel Charles L. Burdett, commander of the First Connecticut Volunteer Infantry Regiment. The alarm became epidemic
after Spanish Admiral Cervera's pitifully small fleet was reported overdue at Havana. Was it cruising the Atlantic coast,

preparing to attack? Bostonians sent their "valuables and securities [inland] to Worcester and to Springfield," wrote Colonel Burdett, "until the safe deposit vaults in those places were taxed to the utmost. . . . [T]he War Department was, by the middle of May, almost filled with petitions for protection and requests of like import."

So "to quiet the fears and stop the annoyance, as much as anything else," the machinery in Washington began to clank and heave, and although the War Department knew by then that Cervera's fleet was actually trapped in Santiago Harbor, Cuba— and shortly word came of Dewey's victory at Manila—the orders rattled out to the field. Colonel Burdett's command was scattered all over the Northeast: one company to the New Hampshire coast, six to Maine, two out to Plum Island in the Sound, and one to Great Gull. Of course, when these troops left for their coastal-defense duties, they carried almost no ammunition. Burdett had asked the Ordnance Department for a supply of ball cartridges, and the Ordnance Department, possibly aware that the entire operation was a fool's errand, dragged its feet. The First Connecticut, a National Guard unit, didn't even have enough ammunition on hand for a decent training program before it was ordered to its various stations; the rounds had to be parceled out like diamonds. When one of its senior officers was forced to resign because of illness, he spent all his back pay buying the regiment four thousand ball cartridges; these were divided equally among the ten companies, meaning that when B Company arrived at Great Gull in June, its men carried ten rounds apiece, and evidently they never got any more during the five weeks they remained on the island.

The company found two hundred and fifty men now at work on the fortifications here—"Italians, Swedes, negroes and a small number of Irishmen and Americans." That's what the colonel said. By June, 1898, there was "but one spot available for the camp, . . . about one hundred and seventy-five feet long by forty feet wide on the south end of the island, and but little above the high tide level. The remainder . . . was excavated as to its entire length. . . . There was no room for any extended order drill, but what there was, was utilized in company and squad drill." Having no Spanish navy to fend off with their rifles and ten rounds per man, B Company set up tight security to protect its postage stamp of a camp from the workmen; one couldn't be too careful with all those non-Americans around. ". . . [N]o one but the contractor or foreman . . . passed the guard line. The workmen and others on the island would gather along the guard line and watch the troops at drill, at guard mount, inspection, and even at play, with great curiosity. . . ." Coexistence was peaceful. Military discipline was good. On one occasion, to be sure, several of the soldiers "took French leave" in a sailboat and spent a few days AWOL in New London; one or two of them then deserted rather than go back to the rocky little camp in the middle of the Sound, but the rest returned. Nothing more serious than that occurred before the company was ordered back to the mainland in mid-July, and the regiment left for a camp in Virginia.

The next year the fort was named—Fort Michie, after the same young officer, Lieutenant Dennis Mahan Michie (killed July 1, 1898, in the fighting around San Juan Hill), for whom West Point later named its football stadium; Dennis Michie had introduced the game of football to the Point when he was a cadet. A small artillery unit moved onto the island while the construction went on. The eastern half of Great Gull, commanding the entrance to the Sound, became dominated by ridges of packed sandy soil—earthworks to catch enemy shells—and by gun emplacements and magazines. At the western end, a village of buildings and towers was erected—barracks, officers' quarters, storage sheds, fire-command posts, a hospital.

For a while the terns still tried to nest on Great Gull, but by 1900 the bulk of the old colony had left. That year a "few essayed to use one of the extreme points for nesting purposes," one observer reported, but the Army's then thirty-man force at

Fort Michie "gave them a warm reception," and "they sought safer and more peaceful quarters." Great Gull was "totally deserted by the terns," because of the Army's presence.

That appears to have been the impression held by the entire ornithological community for the next fifty years. An Army fort means lots of people, right? But in this case it didn't. Not as a regular thing. During World War I, one or two companies of coast artillery were assigned to the fort; then for years afterward the place was usually left to the caretaking of only a few men. Occasionally a full company would be stationed here, but Great Gull was simply too isolated and difficult to supply to be treated as a regular station. That was so even after the early 1920s, when the Army placed the sixteen-inch gun out at the eastern end to protect New York from a "run-by" up the Sound by an enemy fleet. The installation of that gun meant another major construction project, huge earthworks, a great reinforced-cement gunpit and underground cavern back of it—to hold ammunition, the gun crews, and the generators that supplied power to the gun.

This massive structure was built at a cost of about a million dollars, and the gun seems to have been shipped from the Army's proving grounds in Maryland, transferred to a barge at Jersey City, brought out here, derricked onto a railroad car, hauled by locomotive down the island, and set in place—this by the end of 1923. It rested on a "disappearing carriage," which meant that the gun could be lowered into the safety of the emplacement when it was being loaded and would be exposed to enemy fire only when it was being fired itself. The story— perhaps apocryphal—goes that the first test-firing of the gun blew out windows on the north shore of Long Island; Captain Malloy tells me that subsequently the Army took care to warn residents along the nearby coast each time the gun was to be fired, so they could open their windows to let the shock waves through. That didn't happen often. The gun had stood silent for eight years when, in October, 1934, it was used in the first sixteen-inch-gun target practice ever held in the continental United States. The target was more than fifteen miles away, out of sight of the island. The shells wheeled out from the underground cavern and lobbed in the direction of the target weighed more than a ton each, and required more than eight hundred pounds of powder per round. There were three test shots and then six shots for record, all fired within a space of seven and a half minutes.

The gun couldn't have been used much after that; such practice firings were not only a major expense; they must have constituted a considerable problem, requiring the rerouting of shipping to and from New York. And then, as Captain Malloy so succinctly puts it, "they found out everything was flying around up here in the air, and the disappearin' gun wasn't nuthin'." A 1940 memorandum from the Office of the Chief of Coast Artillery to the Ordnance Office of the War Department noted that an Air Corps advisory "indicates that a 16[-inch] permanently-emplaced gun is a relatively easy target from the air" and that "such a target would be a desirable Air Corps mission. . . ." Furthermore, advances in ordnance had quickly made that model of gun obsolete with or without the possibility of air attack, and before the end of World War II the disappearin' gun was classified No Longer Required. Eventually it was taken back down the railroad track to the dock, where the barrel was cut in half and sold for scrap.

I find no mention in the ornithological literature about terns on Great Gull during the period between the building of the fort and the end of the second war. But that may well be because the ornithologists stopped checking. A couple of nonbirders offer testimony that indicates that the birds did come back. One woman, now living in western Connecticut, used to sail on the Sound with her grandfather in the 1930s, and she remembers the place well, because it was "covered with gulls"—for which I think one should read terns; when the Army reoccupied the fort in force for World War II, according to Captain Malloy, the terns behaved as if they owned the place. "They'd fly down—you know—bomb you, and discharge their load on top a you," he says, "soldiers walking around with all that bird manure all over their jackets." Terns do not react so strongly for no reason; they were protecting nests and eggs and young.

There wasn't much room for terns on Great Gull once the Army took over again, Captain Malloy admits. Mostly they "went on the old fort down on Gardiner's Island, and on Cartwright Shoal, and around over in there. But they come while the war was on every day to look at it, to see what was going on. They made better inspections there than the United States government did."

If the terns had in fact returned, the Army and the war outlasted their persistence; by the end of the war there was no tern colony on Great Gull in the summers. Some four hundred men had been stationed here; to shelter them the Army put up unpainted wood structures, and tore down older brick buildings and buried them in the island—to Captain Malloy's lasting disgust. Then the Army left the island fortress, as it did its other coastal-defense forts. Enemy fleets would not be trying to attack American cities when aircraft and rockets could do that so much more efficiently. The guns were removed, never having fired a shot in anger. The wooden buildings were taken down and salvaged. Local wreckers began slipping ashore and quietly stripping the fort of everything that could easily be carried off.

And now began the current chapter of the island's history. In 1948, through the goodness and wisdom of the War Assets Administration, Great Gull became the property of the American Museum of Natural History. The museum had an idea of creating a wildlife research station here, and of restoring the island as a breeding ground for terns, which not only had given up nesting here but were becoming scarce all around Long Island Sound. The Linnaean Society of New York, itself picking up where it had left off in the 1890s, volunteered to assume charge of this project for the museum, and shortly undertook weekend projects to post the island, start studies of birds, plants, and mice (the Gull Island Meadow Mouse was long extinct), clear away vegetation along the shore behind the Army's riprapping, open up nesting territory for the terns and try to attract them back. Plaster-of-Paris tern models were placed in areas where the terns ought by rights to nest, an attempt to decoy that had never worked on terns before and didn't work this time, either.

Two members of the museum staff, Catherine M. Pessino and Lois J. Hussey, spent a month on Great Gull during the early summer of 1950. They watched several hundred terns perch daily on the dock pilings and the outlying rocks; they saw the birds fish along the shore. But not once did a tern land on the island itself or pay any mind whatsoever to the decoys. The Linnaean Society's Great Gull Island Committee kept at the work of opening nesting areas and posting and reposting the shore, but in 1952 the terns still would not come to land. They seem to have been acting out the same tentative approach that is seen each spring here, but in slow motion. Catherine Pessino and Lois Hussey watched them for what must have been a frustrating eight days in June, 1952. "Each evening from about 5 P.M. to sunset," wrote Lois Hussey, "groups of 2 to 17 Common Terns would fly over the island, swoop low over the ground, hover over different areas, and chase other birds from the areas that the terns seemed to examine so closely."

Perhaps one of the reasons the terns were not enthusiastic about Great Gull as a nesting site that year was an absence of baitfish near the island. Pessino and Hussey had it in mind to take ten-day-old tern chicks from a nearby colony, carry them to Great Gull, feed them there until they were able to fly and fend for themselves, and thus create the nucleus of a future colony. But the preparatory experiments in netting fish near shore were unsuccessful, and the two women gave up their scheme.

Meanwhile, Captain Malloy had contracted to dismantle most of the buildings left on the island, to knock down one of the fire-command buildings, and salvage whatever materials he could as his fee. By the summer of 1954 the job was done. He left three brick buildings standing—two officers' quarters and an ordnance shed, which had apparently last been used as a carpenter's shop. All were in good shape and had their heating plants intact. He also left alone several blockhouses and the central watchtowers and the major fortifications.

But Great Gull remained essentially untended except for a few days each year, and it continued to attract wreckers and vandals. All the glass in the windows of the brick buildings was broken. The heating plants and plumbing and wiring were ripped out. Even the copper flashing around the eaves was torn down and boated away. The weather ate at the buildings; rain washed down the inside walls and through the ceilings, crumbling the plaster, rotting the floors.

Meanwhile, nature was working on the island in other ways. The great wharf that had reached out some six hundred feet from the north shore was breached in a bad November storm in 1950 and began to disintegrate. Grass and bayberry and chicory and purselane and poison ivy chewed at the complex of roads and walks, moving inward from the edges and outward from cracks that appeared as the island heaved and shifted gently under the macadam and cement.

At last, in July, 1954, terns landed again on Great Gull, though only a few of them, and they didn't nest. But during the 1955 breeding season, Irwin Alperin, the president of the Linnaean Society, made several inspection flights over the island, and out near the sixteen-inch-gun emplacement at the east end he found a small colony nesting. From that summer on, the terns increased. The island had been theirs to begin with, and it was theirs again. An estimated two hundred pairs of common terns nested at the east end in the summer of 1956, two hundred and fifty in 1959, a thousand terns in 1960 and 1961, three times that in 1964, and by 1964 perhaps as many as half the birds were roseate terns that had evidently been displaced from breeding grounds along the Massachusetts coast.

Through the summer of 1964, relatively little research work was carried out on Great Gull. Mouse studies had been conducted by researchers from Columbia University and the New York Zoological Society; a few bird censuses had been done, and some banding of birds. In 1964 a young ornithologist, Helen Hays, was among the members of the Linnaean Society to visit the island on tern censusing and banding trips; her arrival was to prove momentous for the work here. She was deeply impressed by the possibilities the island offered as a physically limited and isolated study area with a varied but still manageable range of faunal and floral species. She began to urge that the research be intensified. In 1965, members of the museum staff and the society—Helen Hays among them—came out to the island every weekend from early June to late July, and in 1966 there were researchers here from May 20 through June 26, at the peak of the nesting, and for weekends after that through early August. Gradually the studies expanded and grew more complex. The island was marked off into "quadrats" or squares, to give territorial exactness to the studies. Color-banding was begun, to make it possible to identify each banded bird at a distance as an individual, instead of having to trap it and read the number on its aluminum band. The distribution of the two tern species here was investigated; studies were started to learn more about the tenacity of terns in returning to former nesting sites each spring; every tern nest that could be found on Great Gull was given a number and marked with a wooden tongue-depressor stuck in the ground alongside, so that individual nests could be studied throughout the season; the eggs in those nests were numbered in sequence, as they were laid, so that each could be kept track of. Meanwhile, research on other nesting species on the island was started, as well. Publications began to flow from all this work, and Helen Hays, by 1969 well established as the Great Gull Island Project's motive force, was named its director.

The terns flourished in the benign human company of the

natural scientists—though there were more people here every year, and for longer periods of time. They expanded their nesting grounds in 1966 to include the western end, where the old fort hospital had once stood and where, even earlier, B Company of the First Connecticut had briefly bivouacked in the summer of the Spanish-American War. There were about two thousand pairs of common terns in the colony, and some fifteen hundred roseate pairs; a tern-less spot eleven years earlier, Great Gull Island had the largest known roseate tern colony in the western hemisphere.

The persistent study began turning up interesting information. Some of the roseates, for instance, were making their nests in totally unexpected places. They had been known to nest alongside common terns, though they tended to choose more thickly grassed spots, and they did that here; but on Great Gull they were also found nesting in little caves and crevices in the granite riprapping, where the common terns *never* went. No one had seen roseates do that before. And then, in 1972, an astonishing discovery was made: here and there on the island, commons were interbreeding with roseates and producing hybrid terns. No one had seen that before, either. These were the sort of lucky scientific rewards every researcher dreams about.

By now nearly forty articles, papers, and notes on observations have been published, and more are on the way, as a result of this project—on the terns, on red-winged blackbirds, barn swallows, ruddy turnstones, spotted sandpipers. Thus is the note-taking formalized.

The roads, the rubble, the foundations and cement floors of the vanished wood buildings, even the standing brick and cement structures are slowly being swallowed up by vegetation. The great warren of an underground bunker back of the sixteen-inch-gun emplacement is pitchblack inside and almost empty; splinter-dry planking bridges the empty cable and pipe trenches in the floor; urinals rust against the walls; the rusted remains of a stripped-down Model T, which Captain Malloy used in his salvage work, sits in the dark where it was left years ago. Outside, at the center of the huge gunpit, concentric steel-edged circles of steps lead down into a pool of green water where piles of lumber were dumped and now lie gray and sodden, perches for terns and barn swallows and an occasional heron. (A female mallard once led her downy young to swim in this gunpit; no one knows how, because the drop into the pit from the rim is a long one, and the walk through the dark tunnel is two hundred feet. In any case, all the ducklings died in a few days.)

Elsewhere other empty gun emplacements—the old positions of the guns marked by circles of cut bolts in the cement—sprout rust and tufts of grass. Elevators that once raised ammunition from the magazines below are gone or rest rusted in their shafts. All over the island, open manholes (most of their covers having been stolen) lead into small rooms, or less than rooms, where cables met, or into what Captain Malloy calls "resevoys," for the drinking water that had to be brought out from the mainland by boat. The platforms on which generators stood are empty. In the underground bunkers that held powder and shells, in the cement rooms built into the earthworks, where soldiers would have waited out naval bombardment, are scraps of lumber, bits of old brick and crumbled masonry, and drifts of dried grass and bay twigs.

It is a child's dream of a summer place, and if there are ghosts, they are harmless; no nightmares of battle here. A benign fortress, full of play forts and huge iron staples bolted to the cement for ladder rungs; of high places to climb and from which to shout commands; of tunnels and stairways and slit windows over which were printed, in black paint now flaking away, the names of distant points of land to remind the Army's watchers what direction they were looking in.

The Gull Islanders who took the Army's place have added their own such directions: everywhere you look you see evidence of the marking off into squares: letter-and-number combinations painted on the sides of blockhouses, on the corners of the brick buildings, on rocks, on small signs posted in the open. (Great Gull into 174 parts divided is.) And along the shore many of the huge stones have been given numbers, also to help researchers keep precise records of how the birds behave.

There is money enough only for supporting the work itself— and not enough for that, in fact. The little aluminum boat, known as *Bobbin*, which we brought out on the *Sunbeam* this afternoon, is actually only an aluminum sieve. We put it in the water, planning to haul it ashore from there, but it filled with water and was close to swamping before we managed to horse it up on the beach. It seems to me that an island research station ought to have some sort of serviceable work boat, but the project's money doesn't stretch that far. Yet as in other such operations, the members of the project take a certain wry pride in what they can accomplish on a shoestring. "Terns on a dime," as one of the young researchers once put it. So almost all the improvements to the dilapidated buildings have been strictly catch-as-catch-can. The two officers' quarters are now dormitories, strewn with cots and torn mattresses and sleeping bags; the roof of one of them has been retarred recently, eliminating the worst of the leaks there, but the other has not been so favored yet. The inside of the cavernous old carpenter's shop (tongue-in-groove walls and ceiling, windows mostly boarded over) which houses the kitchen—the "G.G.I. Essen Und Fressen Society"— in one room, and various desks, tables and bird-banding equipment in the next two, has been painted a sunny medley of white and bright blue and maroon and yellow since my first visit here several years ago, but the last person to go to bed each night takes care to store things in spots that are not under known leaks in the roof. Furniture has been scrounged from the mainland or fashioned out of stuff cast up on the beach. Lobster traps, for instance, prop up various plank tables. The sheets of plywood that cover empty windows and doors to keep out the weather often have holes cut in them, to let in the light, and over these holes pieces of scratched second-hand plexiglass—many of them dis-cards from American Museum displays—have been fixed.

This crumbling fort, with the sketchy make-do signs of new uses superimposed, should not by all rights exude any sense of order. But it does. The order is imposed partly by the work, the act of research, and by the subjection of almost everything else to that act. It is imposed, too, by the inevitability of nature's re-silience, by the constant encroachment of sweet-smelling bay-berry, beach pea, bindweed, all the plants that grow rank on the island despite man's battlements. And it is imposed, above all, by the birds.

III

April 30: A clear morning, with the wind westerly. All of us are up before the sun. The young bird artist, Tom van't Hof, is out in the net lanes, shaking out the mist nets to catch the migrants that are coming down to land after a long night's northeastward flight. Helen Hays and Grace Cormons stand in the Carpenter's Shop, drinking coffee, impatiently waiting for the light to brighten enough for them to tour the island with telescopes and find out which of the color-banded blackbirds have arrived back.

When they leave, I (still feeling a little like a tourist) make my way up the iron ladder to the third floor of the taller of the two cement watchtowers to see what *I* can see. From that vantage point one can watch most of the island and the surrounding water. To the east, beyond the Carpenter's Shop and the vast cement construction in back of it—the circular pits, tunnels, storage rooms, and elevators of what was once a two-gun battery—beyond the earthworks raised to protect the battery and the Carpenter's Shop in the days when that building was an ordnance shed, the land drops to a spacious meadow. The meadow spreads from beach to beach, out to the the great mound—earthworks topped by a broad cement roof—beneath which is the cavern as big as a football field and on the far side of which is the huge gunpit that the cavern serviced. Beyond the mound I can see Little Gull Light and Block Island Sound. Looking the other way, westward, where most of the fort's buildings stood, the terrain rolls gently out to a narrow point. Some of the Army's cement floors and the macadam roads and walks can still be seen, hemmed in by bayberry and meadow. Out there, Helen and Grace move slowly, telescopes on tripods carried over their shoulders; pause, set up the scopes, peer through them, write notes, pick up scopes, move on. Below me, Tom moves from net lane to net lane, removing birds from the nets, carrying them back to the banding table in the Carpenter's Shop, where he weighs and measures them, puts an aluminum band on each one, comes to a doorway, and releases them. . . .

After breakfast, Helen and Grace set out baited box traps, made of hardware cloth, to catch blackbirds; there are a good many unbanded blackbirds here, along with the old friends from other years. I return to the tower to watch for passing terns and for any marked blackbirds that may be hanging around this part of the island. Tom continues his rounds between the nets and the banding table. As the morning wanes, the netted birds become scarce, and Tom hyphenates his routine by digging in the little vegetable garden at the foot of the tower. After a while I climb down and take over the job of checking the nets. Tom patiently teaches me how to unsnarl birds from the fine mesh, how to measure and weigh them, how to put on the bands and fill out the record sheet. By early afternoon I have at last denetted, measured, weighed, banded, and released the first bird I have ever processed without help—a male rufous-sided towhee.

Late in the afternoon, Helen and Grace return from a tour of the blackbird traps carrying a red-wing that is here for its seventh year, which makes it the oldest red-wing in Great Gull's records. They are delighted with the discovery, not only because of the new longevity record but also because they know a good deal of that particular bird's history. Now they remove the old four-band combination, which has become worn, and give the bird a new set before letting it go.

May 1: One of the most remarkable aspects of this project is the way it attracts—and depends on—such young bird students as Tom van't Hof. Tom is one of those bright, dedicated teen-agers who constantly amaze me by how much they know and how completely they have been overtaken by birds. In many aspects of ornithology I—well more than twice his age—can be only a student to this eighteen-year-old, who is not even officially graduated yet from his high school on Long Island. He tells stories about his school career that lead me to believe he has had very understanding teachers and school administrators. He cut classes more or less at will in order to go birding, sometimes whole days; or he would arrive in school at midday, for example, soaking wet and stinking with marsh mud from his hips down, having just spent five hours in the rain chasing a warbler he wanted to photograph. He is a good student, earnest and intense, and his teachers evidently just learned to put up with his eccentric behavior and, when they could, somehow bent his enthusiasm back into the Course of Study. The reason he's not fully graduated yet has something to do with a half credit he needed in English; so he suggested he write a scientific paper for that half credit, and the school agreed. His paper was about an esoteric and complicated study he did out here last summer, measuring the temperatures of tern eggs in the nest. Not your common garden-variety piece of work for high school English, though God knows there ought to be more such flexibility in secondary-school education. Most of his classmates will have spent this spring hovering over the mailbox, frantic to find out if and where they will go to college; but not Tom. Having finished his paper, he decided to fulfill a wish he had had for years, to stay by himself on an island; he came out to Great Gull in the first week of April, and he has been banding birds and painting birds—mostly heads, some very good—and making the island ready for the rest of us. If no college accepted him, he was prepared to spend a whole year here, but as it happens the mail we brought included a welcoming communiqué from the University of Michigan's School of Natural Resources. He is tall and lean, with a handsome face and perfect teeth, black hair, black beard; he's shy, smiles easily, smokes Bugler tobacco in roll-your-owns, memorizes the funnier radio ads and recites them in an abashed, rushed, pleased manner.

We worked on the project's vegetable garden together today, then went out to a bluff overlooking the eastern meadow and built a bird-attracting brush pile along an alleyway cut earlier

through bayberry bushes for the stringing of a mist net. It had been a fabulous day for birds; they had poured down with a light rain early in the morning—hundreds of white-throated sparrows, swamp sparrows, chipping sparrows, kinglets, warblers, a rose-breasted grosbeak, a northern oriole. All over the island you could hear birds lisping in the underbrush and scratching in the leaves for food. As we worked on the net lane Tom suddenly remarked that this was the way to live—birding all day, doing jobs like this, planting your own garden. For him, Great Gull is a place to plug in to life, to help create, not just to follow, a meaningful routine; a place for discovery. There are dozens more of the same bent and youthfulness who have been thus nourished here and who have at the same time fed back into the project their enthusiasm and energy and talents.

May 3: Rain and cold all day. Much of the day's activities took place in the Carpenter's Shop "office" next to the kitchen. It's often used as a banding room, but it is also a place where files are kept and writing done. Helen has her desk there, beside a plexiglass-covered hole in a boarded window, and she was at her typewriter most of the time, polishing the final draft of a Gull Island report. Grace was at another desk, working on records. In an oil-drum stove in one corner, burning driftwood crackled noisily.

This morning Tom found a field sparrow that couldn't fly, and it was brought inside, out of the rain, put in a box with a screen for a lid, given food and water; by midafternoon it was strong enough to scramble out of the box when someone lifted the lid to check on it, and it led us a fine chase around the room, though it will probably need some time to recover fully from whatever ails it.

I spent my day mostly in my room in one of the officers' quarters, putting it in some kind of order: clothesline hung, desk built from lobster traps and a piece of wood found on the beach, holes cut in the plywood sheets over the windows and plexiglass pieces "installed." As I sat at my desk this afternoon, surveying the many wonders I had performed, my breath steamed and the roof leaked rainwater into the former bathroom next door.

May 5: The wind peals down from Canada this morning; it gusts and buffets across New York state and western New England, reaches the Sound and flattens out with room to run, kicking up whitecaps as it goes. The rain has been blown from under the blue sky, and the air is brilliant with the glitter of the sea.

This shining northwest morning I hide from the wind below the bunker near the dock, and do some tern farming. We have been at this business for several days, on and off. The beach grasses and yarrow and bindweed and beach pea and poison ivy and bayberry grow more luxuriant each year, and creep as far down the beach as high tide will let them, gradually pushing the terns out of their old nest sites. When there were dozens of suitable breeding places on the Sound, that made no difference. Birds could abandon Great Gull for something better, and eventually one of the lighthouse keepers would burn over the island, or he would pasture a few cows here, and *they* would trim the place. Or an autumn hurricane would career down the Sound, spilling the sea high on the beach, perhaps even sweeping salt water clear across the island in places. The waves would scour away the plants from the upper beach or bury them in sand, and when the water receded it left behind salt in the soil to retard the inevitable return of the vegetation. Next spring, there was more room on Great Gull for the terns.

But there are no cows on the island now. And Great Gull is not so susceptible to the force of storms; the Army saw to that with its riprapping and its seawalls—which also destroyed acres of beach. The buttressing isn't absolute; * but it's sufficient. The plants take over. And now there are few alternatives for the

* Captain Malloy remembers a "breeze" that left water standing several feet deep in the middle of the island and ruined the Model-T he had left here.

terns to choose; man has cottaged and public-parked and marina'd and dumped almost everywhere. So on Great Gull, like the Austins on Tern Island, we tern farm.

The first restoring crews out here after the Army decamped—volunteers from the museum and from the Linnaean Society—tried spreading Borax; they tried a herbicide; grass has been burned off or covered with sand; here and there around the island are tattered sheets of black plastic, stretched over grass in hopes of smothering it. But the main approach is to dig and hoe in the rocky, rubble-filled soil and pull at the plants by hand. (Tom, who handles a hoe with great energy and effect, has composed a litany for our tern farming, a work-chant: "Chink-chink-UNH! Chink-chink-UNH!")

In the long run, the vegetation is no more defeated by our labors than it is by the Army's works. The plants require only a month, a season, at most a year, to reclaim their territory along the shore.

As I sit alone on a beached piece of foam rubber and pull at the incredibly long and wandering roots of what I take to be bindweed, the common terns fly overhead, demanding—it sounds to me—"Hurry-hurry-hurry-hurry-hurry." *Well, you, too, fellows,* I think. They have not yet made their first touchings-down, though there are more of them scouting us each day. We expect they will land shortly.

This tern farming, if it does what we hope it will, imposes on us the requirement to do more and more of it every year. We are increasing the colony's productivity, thus making it larger, thus demanding that we clear more of the island each year. If we do *not,* we will have played the terns a dirty trick. This is ironically like the multiplying problems the United States forces on itself and on the rest of the world with its Green-Revolution agriculture and its immense shipments of food to nations that have outgrown their food supplies. Where is the logic in feeding such nations and encouraging even more serious overpopulation, which someone will have to feed, house, educate, supply—and which the earth must bear? And where is the humanity in withholding that food? Impossible.

Sitting here, tugging at roots, I know that I'm not getting them all out. Even if I were, the plants farther up the beach would quickly invade the territory I'm opening. This will all be covered with greenery again in a few months. I regard tern farming the way one does doing dishes or making beds, other jobs that are never finished when finished.

I have the feeling that we environmentalists are all grubbing away like this, earnestly clearing spaces that will be overgrown in a season. How hard we are working to change things. But I sense that we are going about it in routine ways—and hurriedly (there is so much to do)—when what is needed is something cataclysmic on our part, something earthshaking. Likely the earth will shake itself before we get a good grip on its lapels. We are urged to work within the system for its salvation, and most of us agree wholeheartedly. I know it's the approach *I* feel most comfortable with, even if the system allows us few solid handholds. Yet I read environmentalists who say they see things turning around, and I think, *Whistling in the dark*. Simply the relationship of our energy needs to the production of food to the growth of population should have us all behaving not just like Jeremiahs, Isaiahs, crying Woe and Shame, but like Moses, leading.

And where, pray, shall we lead? There are no geographical escape hatches. The Red Sea will not part for us. Apollo offers no tourist fares. Confined to the planet, we can only change ourselves, and the change (such as it is) seems to tend the wrong way. The greater the population, the greater the necessity for governing it, the less our freedom. Where else can we go? We have become greedy, knowing destroyers of rivers and oceans and farmlands and plains and forests; terrorist avengers turning technology against the general weal (skyjacking, kidnapping, letter-bombing, assassinating, making war, poisoning other men's harvests, other men's air and water), and much of this is

the result of pressure, the pressure of people and technology. Yet I speak of control, and who is to do it? Could the rationale and energy of environmentalism repress in the same degree the other savior *isms* have? I doubt not. But the world has *already* entered the starving time—

I weed peacefully and hopelessly, hurling clumps of bindweed up into the wind and watching them be blown to the east; I tear out rocks to get at the roots that cluster beneath, throw the rocks over my shoulder, move on from each cleared square foot to a new patch. I pause to take off my gloves, smell the sea, smoke a cigarette, and with my binoculars check the colored bands on the legs of a song sparrow that has dived over the bunker above me and now perches briefly on broken cement. I set the glasses down, put on the gloves again, and focus minutely down on the offending plants.

May 11: I am spending the last hours of daylight at one of the old blockhouses near the docks. I sit outside, facing the Sound, with my back against the blockhouse wall, on which an artist from the Linnaean Society has painted a large sign, decorated with pictures of terns, warning RESEARCH STATION DO NOT LAND. My assignment this evening is to record the "fly-outs" I can see from time to time on either side of me. The terns continue to behave toward the island like lovers who are not sure of the response they will get. The first bird to touch down nearby, about a week ago, landed on the pilings of the long dock and then took off. Then this week, while I was back in Connecticut, terns landed on the island itself for the first time this year. The next milestone would be the first night they spent ashore, and last evening Helen Hays led a sizable party of us (the number of people here has grown) out to North Beach; in the dark we walked toward the tooting from Little Gull and the God's-eye of its light, then shone a beam of our own down onto the beach, to discover terns roosting there.

So they're settling in. But they're still uneasy. A lot of them line up on the pilings in front of me and rest there; perhaps some of them haven't even touched land yet. And many of those on shore appear fidgety. Every now and then their Babel of cries will cease, and great drifts of them will rise from the breeding territories to fly silently out over the water, circle, and fly back, calling at last as they settle down again. There are several names for these fly-outs, and apparently several different kinds: *alarms*, *dreads*, and *panics*. You see them throughout the breeding season, but never so frequently as now. There are sometimes obvious reasons for fly-outs: a predator appears—a peregrine falcon or a marsh hawk or a short-eared owl; but in most of those cases the fly-out will quickly become a general and noisy attack on the intruder. Ralph S. Palmer, who studied common terns in Maine in the 1930s, noted that it sometimes took only the tooting of a boat's horn offshore to put every bird up in a panic. Even a snap of the fingers can do it. And many times, here, we simply can't see or hear anything that might have caused the fly-out. Helen Hays thinks the birds may be reacting to something we can't perceive so acutely as they, such as a change of light. So I watch for light changes this evening—the sun sliding down behind a cloud, appearing again, going behind clouds again. I can't see any correlation. The fly-out seems to begin with a few birds and quickly spreads through their near neighbors, until hundreds of terns are in the air. But there are apparently certain limits to the influence of the birds on each other. A flight out at the west end of the island may or may not be echoed by one at the east end, or by the birds on the beach on either side of me; and if it does seem to be echoed, thirty seconds or a minute may pass before that happens. If the birds from the west end fly out and then circle back over the dock, the terns sitting there may react, and they may not. How much of a fly-out takes place may depend— no one knows—on the kind of fear the birds communicate to each other, either silently or in the tone of voice of those who stay put and do not join the fly-out. Again, we watch for a *sign*, and make notes about what we see and hear, hoping that hundreds of such notes, read together, will eventually reveal the answer.

Between fly-outs, the terns are courting. I have always thought that they were beautiful at rest and beautiful in flight, but courting, they move into another dimension of grace. Here, flying in, is a common tern carrying a small, silvery fish cross-wise in its bill. As it descends, a sitting bird stands up and begins to call, *ki-ki-ki-ki-ki*, turning with mincing steps on car-

mine feet to face the first bird as it passes overhead and circles and then lands a few feet away. The new arrival waits there with its fish; the other tern holds its place, too, and cries to the other, apparently begging. The fishbearer finally struts across the space between them, muttering *kor-kor-kor-kor* in a low tone, and the fish is snatched away and swallowed. Now they stand almost shoulder to shoulder, facing in opposite directions, past each other; their long, white tails are cocked up, their wings held away and down from their bodies like white-trimmed gray cloaks, their necks stretched, red bills pointing up—elegant, stylized sculptures. Now they dance in tiny steps, a stiff, formal circle dance that involves bows of the head and twistings of the neck. Now they pause, tails still erect, wings drooped, heads up—and suddenly one of the two takes off, and the other follows.

They rise into a sky filled with hundreds of terns in various patterns of courtship flights—the ground dance continued. I try to watch what my pair does, but I lose track of them very quickly. Overhead, birds beat in great arcs, in twos and threes and fours, carrying fish or chasing birds carrying fish, heads bent, growling *kor-kor-kor*, heads held straight, calling *ki-ki-ki-ki*. Some are too far up to be heard above the general din, if indeed they are calling. They follow great circles; up there is one bird in a pair well behind the other, yet not cutting across the circle to catch up; the pair rises higher and higher, and the flight continues like that until my neck aches and my eyes are dazzled by the distance. Now I lose that pair, but here, coming down, are two other birds in a long glide, one flying only a foot or so higher than the other. They swing apart, one to one side, one to the other, and then they swing back, pass each other, swing apart the other way, then back, past each other, and back, describing between them diamond patterns in their descending.

Other terns are fluttering up, wings beating very fast—sometimes alone, sometimes in what appears to be a contest with another bird. Dozens are flying straight courses to and from fishing, dozens more chasing them for the fish they carry. Constant motion, constant sound: *Eee-ah, eee-ah, churry, churry, eee-ah, ki-ki-ki-ki-ki-ki, kor-kor-kor-kor*. Some people seem to have made some sense of it, watched it carefully, made notes, sketched and photographed, and then sorted things out. I can't. I've never done this before. It's a new country, with thousands of strange inhabitants, and I understand neither the language nor the customs.

Eventually the posturing and the displays will lead to a mounting, when one bird will stand on the other's back, sometimes for minutes at a stretch, and in that position they may grapple beaks; not exactly billing and cooing—more like beaking and screaming. If one or the other of them is not ready to proceed, the top bird will eventually get down. But if the signs are all propitious, they will copulate. The genitals of both male and female are on the underside of the base of the tail, so copulation in that mounted position requires a twisting of tails to bring the genitals in contact, and there is a lot of wing flapping and struggling to hold position and balance.

It is only during copulation that we can be sure of the sex of noncaptive terns. The male is always on top. Otherwise, unless one can see colored band combinations or the birds have been boldly marked in some other fashion by researchers, there is no

way to tell the sexes apart by sight. That presents a major problem to the observer who is trying to discover the rules of tern courtship.

The terns have developed patterns of gestures and calls that have as a result the perpetuation of their kind. The ground-and-sky dance is an example. The creation and maintenance of territory around nests is also carried out with gestures and voices, mostly harmless and ultimately pacifying behavior—as we, by shaking hands and saying "How are you?" to an adversary, or sometimes by showing the flag or the artillery, make a kind of peace. So we and the terns have something in common—bonds, if you will. And the physical characteristics we have in common are miraculous, when you think of them: nervous system, heart, lungs, stomach, intestine, liver, kidneys, the proximity of anus, ureter, and genitals, etc., etc. Either we sprang from the same cell, or the Creator was a mighty unimaginative artist.

"The Creator." How easy it is to slip into that perspective. "The Creator," "The Lord only knows," "God will provide," "going to meet my Maker," "we are all children of God." Really, we are a spiritually primitive race.

Explorations into first causes are frustrating and unsettling, and the simple magic represented by the idea of God or gods, and all the accompanying paraphernalia, is a comfortable catchall. But what that magic does, in effect, is take the responsibility for our welfare and the welfare of our planet—not just the responsibility for the fact of our existence—off our own shoulders. It permits and encourages a spiritual *laissez-faire*. The Day of Judgment seems to me a marvelous cop-out, the ultimate self-fulfilling prophecy, the excuse not to concern oneself with any responsibility to the future world of the living (as opposed to the harp-playing) child or grandchild or great-grandchild. Our species plummets toward the Last Trump (and I am convinced that it does) with all flags flying, crying Hooray.

IV

May 14: Clear morning, wind westerly and not southerly, so the expected major wave of migrating land birds did not take place, although a few new birds have arrived. We spent the day banding those that were caught in the mist nets; we did some tern farming in territory the terns haven't found interesting so far, set up a few white canvas blinds for observation, and searched the island for the first egg and other signs of nesting.

A tern usually begins creating its nest by settling its breast into the sand or the drifted dry grass or seaweed and kicking material out behind it, pivoting slowly in a circle and forming a hollow. That's not always the case; sometimes, for instance, one of them finds a natural depression—say, on the top of a rock—

and uses that. Often a certain amount of grass or pebbles or sticks is added to the hollow, but the birds are very individual about such elaboration and may dispense with it entirely. (Let that stand as a warning of the complexities of tern behavior. There are no neat nutshells here.)

At this stage of the season, a tern is likely to make, or at least to start, quite a few hollowed-out scrapes before settling on its choice for a nest, and all over Great Gull today we found such feather-brushed depressions. No one discovered any eggs.

I suppose it should come as no surprise, but no matter how busy one is, or how hard one is working, this place is *peaceful*. And regardless of the clamor of the birds, and the inexplicableness (to me) of much of their behavior, not to mention the casualness of our own housekeeping and our primitive, make-do accommodations, I feel around myself, around us all, a sure sense that everything essential is in its place and operating as it is bound to.

Helen took me out to the east meadow this evening, to show me the black tulips that bloom each spring amid the rubble of an old Army structure. On the way we discussed the mechanics of a project we have cooked up between us, to which Helen has decided to commit a rather sizable fraction of this year's Gull Island budget. Beginning several weeks from now, I will photograph hundreds of completed clutches of tern eggs, so that variations in shape and size and color can be studied at leisure, in the American Museum, without the eggs themselves having been collected.

She is quite a remarkable person, our director—juggler of disparate talents, fund-raiser, chief scientist-in-residence, author or co-author of many of the scientific papers produced from the work here. Keen-eyed and fair-haired, she seems at first meeting, and in repose, to be a cheery, easygoing type, but she is shy, sensitive, intense, completely committed to Gull Island, with a firm idea of how the station has to operate to survive and carry on its work. Captain Malloy's high compliment (though a backhanded Yankee compliment) to this purposeful young woman is that the terns, who "ain't afraid a nobody, . . . ain't even afraid a Hays."

She was raised in the glove-making city of Johnstown, New York. "As a child," she said once, "I collected everything. Not terribly scientifically. I never liked to *label* anything, but I collected, and my mother was bothered that it wasn't *labeled*, you see, that I just collected it. It was all in the *adding*. When we were in our teens, my brother was building a telescope, and he was interested in stars and all sorts of things, and I was just not too interested in anything. I think my parents thought I was slacking off. When I was a junior in college, at Wellesley, they wondered what I was going to do in life. And so I said, well, I thought I'd do field work." Her parents were skeptical. Did Helen really think she'd *like* field work? Did she even know what it involved? Didn't she, in fact, think it might be a good idea to *do* some of it before she headed off in that direction? "Well, I didn't really like being *urged* that way, but I happened to meet the town librarian, and she gave me a book called *North to the Yukon*, by Florence P. Jacques. I read the book, and in it Mrs. Jacques mentions a waterfowl research station in Manitoba, called Delta, directed by one Al Hochbaum. So I took the book home and said I thought that I'd apply at Delta and see if I could work there during the summer of my junior year.

"I wrote H. Albert Hochbaum. He answered, and the reply was rather cryptic. It went something like this: 'We've never had an undergraduate here before. We've never had a single female here before. Now we have one. Signed, H. Albert Hochbaum.' So of course I thought it meant I was to come ahead. I enthusiastically wrote him and said that I'd be getting out of school on such and such a date, and I'd take a train and arrive in Portage la Prairie at such and such a time. Well, apparently he announced to the group at Delta that there wasn't anything he could do about this, but this girl was coming from the East on such and such a date, and everyone would just have to make the best of it."

Despite that unpromising start, she spent a good summer, much of it assisting other researchers; she was also given a project of her own, to determine the incubation period of ruddy ducks, but it took her most of the season, plunging around the marsh, to find a ruddy duck nest, "and then I watched it so carefully the duck deserted, so I brought the eggs into the hatchery and hatched them there and raised them, and that was the project for the summer."

But she was snared, just as beginners are snared who come *here* with no more solid motives than she had that spring, arriving at Delta. After graduation from Wellesley, she entered the Cornell School of Agriculture for ornithological studies, and the next summer, steered by the momentum of her first experience, she went off to the pothole country of Minnedosa in Manitoba, where ruddy duck nests were more easy to find than they had been at Delta, and took up her work on those handsome little ducks in earnest. That was to command much of her attention for four years; then she spent three years studying glossy ibises nesting in the New York City area, just after those big, dark, sickle-billed birds were first discovered breeding there. These days, Helen sometimes shrugs off the importance of her duck and ibis research, because she published nothing about it. But in the process of doing it she was refining field-work skills and developing her natural bent for careful attention to detail over extended periods that would pay dividends at Great Gull.

This is now a full-time job for her: the field work, the raising of the money to keep the project afloat, the recruiting of volunteers, the cataloguing of the data generated on the island (a job that goes on at the American Museum, in a small, cluttered office known as the Western Egg Room). The work itself and the increasingly impressive results must be a satisfying reward. That and the other ripples she creates in the ornithological community as the young Gull Islanders she helps to train begin to publish. One of the country's leading museum ornithologists calls her a "superb teacher."

We found the clump of black tulips poking up through a jumble of old bricks, as they do each year. That's a nice sign of domesticity, like the Russian olive tree that grows out on the west end and the apple trees planted around the officers' quarters—evidence of man's impulse to pretty up, to make or restore order (tern farming, for instance, gives a co-planetarian its rights to order).

From the black tulips we went to the remains of a building left over, I imagine, from the fort's earliest days, a roofed wooden bunker set behind earthworks near the north shore. Barn swallows are nesting on the beams of what's left, and one nest now holds a single egg. Helen took down the egg and held it in the palm of her hand for me to photograph—a test of the trueness of the color film I intend to use. The egg is no bigger than a fingernail. When I was done, she numbered the egg with a felt marking pen and put it back in the nest.

At 10:30 P.M. the wind is southwesterly and blowing hard. If this lovely half-a-gale is blowing to our south as well, birds smaller than half my hand have taken off from Virginia this dusk and will be in New England in the morning.

How that thought expands the awareness of space; Great Gull Island is indeed a mere dot of a twisted pickle at the entrance to the Sound.

May 17: There is a thick fog on the Sound this morning. I have made my way by flashlight through the underground cavern back of the sixteen-inch-gun emplacement at the east end of the island. At the top of a narrow stairway, where the walls and steps now drip with accumulated fog, a cement-walled, cement-roofed room barely pokes its slitted eyes above the level of the cavern's thick roof. As I mounted the stairs, the echoing sound of my pants legs rubbing together startled me with its loudness.

I have come to read bands, as they say here. I ease my binoculars and my telescope through a crack in the burlap that has been nailed over the empty windows to screen us from the terns, and I scan the birds sitting on the roof in front of me, looking for the Gull Island four-band combinations. Each one I spot, I "read," and note it in my journal. As:

A/B X W/BG
O/G X A/RG
A/GY X B/W

Which, spelled out, means:

The Fish and Wildlife Service's aluminum band over blue on the left leg, white over blue-green stripe on the right.

Orange over green, aluminum over red-green stripe.

Aluminum over green-yellow stripe, blue over white.

Each bird so marked, if it has been returning to the island for several years, has a written record kept for it that is likely to contain a number of different observations about it—when and where it was seen, what it was doing each time, anything unusual about its plumage, and so on; if it has been retrapped since it was first banded, then it has probably been weighed and measured at least once, and that data has been recorded.

There are not many terns ashore this morning—gone fishing, I expect, off someplace in the fog—and the action is slow on the cavern roof. The horn on Little Gull hoots through the fog once every twelve or thirteen seconds. I am struck by its mellowness. Although it is half a mile away, it sounds far softer than foghorns I've heard at twice the distance. And while I am trying to read bands, a special set of atmospheric circumstances develops; I am suddenly in the middle of a great sounding box, as on a stringed instrument; for the full pause between each toot of the horn, an echo reverberates softly in the distance all around.

Sound behaves very strangely. You'd think that if you wanted to be sure to warn ships off a reef in a fog or at night, all you'd have to do is signal the reef's location with some sort of regular noise. Not at all. Many things have been tried, to produce sound effectively on water. The first fog signals in this country were cannons. Bells, of course, have often been used. But the best kind of sound lasts a while and is literally penetrating, so a horn blast is preferable, though far from perfect. Little Gull Light was equipped with a fog bell of some sort in 1829 or 1830, when Fred Chase was keeper. In the 1850s, the lighthouse service experimented with horse-driven compressed-air horns at Beavertail Light on Narragansett Bay and at Little Gull. A horse was harnessed to a wheel, which drove the compressor, and this produced one blast every three minutes. The horse had to work at the job for as long as the fog lasted—which might be days at a stretch. After a year or two of trials, horses were retired from the foghorn business, and another fog bell was brought to Little Gull, to be rung by the keeper and his assistant for as long as a fog lasted. Twenty years later, two steam-driven sirens, or "siren-trumpets," were installed out there, the extra one being a back-up, and during the 1870s the lighthouse service studied the patterns of sound produced by the siren-trumpet around the island. Curiously, one of the phenomena investigated then was the echo, probably the same thing I'm hearing today, a hundred years later. The experimenters noted that the interval between "the blast of the siren-trumpet and the commencement of the echo was very brief; so short, indeed, that the ending of the one and the beginning of the other were generally difficult to distinguish. . . . [and] the slight echo which was heard came from all points of the horizon." That was under less than ideal conditions, the keeper told them. "He thinks [the echoes] are heard most distinctly during a perfect calm . . ."—which is what we have this morning.

The lighthouse service was running these experiments because foghorns did not always do the job expected of them. A perfect example of that was provided ninety-three years ago practically to the day—May 12, 1881. The Sound steamer *Galatea*, bound to New York from Providence, in thick fog and flat calm at midnight, ran aground on Little Gull Island and couldn't get clear for two days. Evidently the *Galatea*'s skipper was rather put out. "It was, as usual, alleged," calmly noted the chief clerk of the United States Light-House Board, "that the fog-signal . . . was not in operation at the time of the accident, and the Light-House Board, also as usual, immediately ordered an investigation." Sometimes the keepers were at fault in such situations, sometimes not. Even today, Captain John regales his passengers aboard the *Sunbeam* with stories about the present keepers of Little Gull Light—Coast Guardsmen—and what he regards as their casual operation of the horn.

An assistant inspector of the lighthouse district was assigned to the investigation. He took testimony from the keepers at Little Gull and at other lighthouses in the vicinity; he talked to the officers of the *Galatea* and of various other vessels that were in the neighborhood that night; he also interviewed other reliable witnesses; and "he reached the conclusion that the fog-signal was sounding at the time of the accident; and that, although the fog-signal was heard at Mystic, 15 miles distant in another direction, and although it was heard on a steam-tug a mile beyond the *Galatea*, that it was heard faintly, if at all, on that vessel; and if heard at all, was so heard as to be misleading, though [when she went aground] the *Galatea* was but one eighth of a mile from the source of the sound."

The assistant inspector spent two days on a boat, going round and round Little Gull Island in various kinds of weather, and "he found the aberrations in audition . . . numerous and . . . eccentric." The chief clerk of the Light-House Board made his own investigation a month later, and he concurred. One clear day, his party found a "circle of silence" where the horn couldn't be heard at all, and he decided to check that out at the end of the afternoon. He and a colleague rowed toward the eastern end of Great Gull, "the siren sounding meantime with ear-splitting force. When about 600 yards away we suddenly lost the sound as completely as if the signal had been stopped. Pulling toward the steamer, not more than 200 yards, we reached a position at right angles to the axis of the siren's trumpet, when we suddenly heard the sound again at its full force."

For "the consideration of those who use these fog-signals overmuch as a guide for their ships," the chief clerk quoted the opinion of the assistant inspector who had made the original investigation, to wit, that navigators had to realize they were not likely to hear a foghorn if it were downwind of them any distance; there "is nearly always a sector of about 120° to windward of the signal"—one-third of the compass rose—"in which it either can not be heard at all or in which it is but faintly heard." Fog didn't seem to make any difference in the carrying power of a horn, but strangely enough falling snow did, and apparently countered the wind effect. "It seems to be well established by numerous observations," said the assistant inspector, "that on our own northern Atlantic coast the best possible circumstances for hearing a fog-signal are in a northeast snowstorm . . . with the observer [upwind] of the signal."

None of this, of course, dealt directly with one of the oddities of the *Galatea* accident: she had missed the steam siren on Little Gull in a dead calm. Possibly it was with this in mind that the chief clerk suggested a list of Don'ts for the mariner, a list that might be summed up in the dictum, Don't Trust Your Ears.

Captain Malloy: "Oh, many times I've been out there, lookin' for Gull Island in the fog, no horn blowin' for a while. Pretty soon you'd hear it startin' in, after we'd been there, tied up. But one thing meant the most to me, I used to leave Noank in the morning with a load of sand, a load of gravel, or whatever, to go to Fort Michie or Fort Terry, but we'd always have something to throw off at Fort Michie. There was one time for something like eight days, one of the Army's own boats never went to Fort

Michie, and that was due to a couple of accidents that had been had. The skippers of the boats were all called in and told, 'Any more accidents, you're responsible; you're gonna pay for'm.' And consequently, when it was too foggy, or something like that, they never made the stop to Fort Michie. If I remember correctly, I think it was *nine* days that I took the bread from Fort Wright and put it on Fort Michie in the fog. And we used to get a reasonable distance off Fort Michie—around between six and seven in the mornin'—we'd stop, have to stop the engines, to get the sound. Pretty soon you'd hear 'em: *'Huup! Huup! Huup! Huup!'* And you'd just drop her right down, there was the corner of the dock, you'd come right to it. And that happened many times, that way."

I like Captain Malloy's stories, their rhythm and pace, their detail. Much of his discourse involves yarns, and you mention a subject and he is good for up to an hour or more, counting the side excursions. He's eighty-three and a half now—a tall, big-boned man, gently stooped and aged-spotted—and he's been doing things that make good yarns ever since he was a child. He went to sea as a cabin boy at nine, and he's been a fisherman and oysterman and special revenue agent during Prohibition; he had his own oyster shop in New London, when the oystering was good; he's ferried goods and people all over the Sound and east to the Cape. Now he's beached. He was ill for a while, and then the engine on his boat gave up the ghost. But to hear him talk, he doesn't intend to stay ashore forever. He's looking for a new engine; he has some oyster beds that need tending.

With or without working foghorns, the Sound can be a nasty place to navigate. Off to the north here is the Race, where the tides, particularly in close to Great Gull, tear along at speeds up to five knots. All over the Sound you find these rips; combined with winds and storms, they have been eating at the land in places ever since it was laid down, and building it up in others. In *The Geology of the First District*, a volume of the 1843 *Natural History of New York*, a writer remarked that the

> eastern parts of Gardiner's and Plum Islands, which are composed of loose materials, are washing away in consequence of the very strong tidal currents and the heavy sea rolling in upon their shores from the open ocean. The action upon these coasts is so rapid as to attract the attention of the inhabitants; and calculations even have been made, as to the time that will probably elapse before they will have disappeared. Rocks . . . that have formed a part of Plum Island, may now be observed at low water, a mile or more from the present shore. . . . Oyster-pond point [Orient Point] is wearing away rapidly by the combined action of the waves during the heavy northeast storms, and the strong tidal current that flows with great velocity through Plum Gut.

Captain Malloy talks about Gardiners Point, on which Fort Tyler was built. "When I first come around here, say in the year 1900, that bar was solid to the old fort from Gardiners Island. It was all full a cedar trees and cedar stumps. Today, there's water across most of it, and the old fort is standin' out there all alone by itself." There also used to be a bad reef somewhere between Fort Terry on Plum Island and the Gulls, he says, but "through 'thirty-eight hurricane and a few breezes, where that we couldn't go in them days, we got 18, 19, 20 feet of water nowadays."

That must have happened out here, too, and split one island into the present two. I am aware that two islands as close together as Great and Little Gull might have gone by a joint name at one time; chartmakers and sailors used not to be so fussy as now. Indeed, "Gull Island" has been used interchangeably even in this century; whichever of the two islands was the focus of attention got the name. But I am still fascinated by the reference to a single Gull Island in the grant of the manor to Samuel Wyllys, a reference that was repeated in the deed to Joseph Dudley. Could the splitting off of Little Gull have taken place only two or three hundred years ago? Long before then, the rips and tides and storms had probably separated the island or islands from the mainland. There's suggestive evidence of that, not only in the very existence of the rips and tides and storms, and what they did to Little Gull in the early 1800s, but also in the discovery of the "Great Gull Island Meadow Mouse" out here in 1888—a mouse so different from its family on the mainland and Long Island that it was considered a new species.

The Devil's Belt, indeed—particularly in the days before sonar and radar. Captain Malloy again—one of his stories that begins at the beginning: "Sergeant Reynolds was the man in charge at Fort Terry on Plum Island after World War I. There was no activity much; he had about eighteen men that took care of the guns and the stuff that was left on the island. And of course the sergeant, with people that had nothing to do with the Army, if you stopped in Fort Terry, he'd always give you a big welcome, and 'Come up and have a beer.' Well, in those days, I drank beer, but since the country went dry and the beer come back, I don't drink any of that. It's too green; it'll kill you too quick. The home brew, I wouldn't guarantee it one way or the other, but Reynolds, he spent most of his time makin' it. Well, he had an old farmhouse that was right close to where the harbor was, and this night—foggy night—we stopped there, and his wife—a little bit of a skinny woman—she was alive the last I knew—she come from the old farmhouse and says, 'You boys better come over and have somethin' to eat with us.' 'Well,' I says, 'We've already had our supper before we got in the harbor here, but thank you.' 'Well, come over, help the sergeant drink his home brew.' So we were settin' at the table, playin' cards. All at once we'd hear *WOOOOooo*"—he did a more than passable imitation, very loud—"steamboat whistle, comin' down the Sound. Mrs. Reynolds says, 'Gosh, that guy is closer in shore than we ever have 'em. And,' she says, 'if we don't look out, he gonna cut the side right off the island.' Well, he didn't touch Plum Island. But he hit Fort Michie straight where that dock was. *Priscilla* was the name of that boat, if I'm right. *Priscilla*, yes. *Priscilla* hit that dock right head on, she just split that dock right out. Her bow, before they could get her headway stopped, almost touched the rocks on the shore underneath the dock. She spread that dock right across—twelve-by-twelves, fourteen-by-fourteens, she just cut them, went right through them."

I have found the story of the *Priscilla* elsewhere, too. A side-wheeler on the Fall River Line, she had been running between New York and New England for years when on the foggy night of May 31, 1926, she came a cropper. Navigating by sound and by the clock, zigzagging from mark to mark, her master had her on a southeast course at the time, headed for Little Gull Light, as he calculated; when the horn at Little Gull was heard, he ordered a course change to the northeast, to carry the ship past the Gulls, which he reckoned from the clock were still some distance away. But before the turn was completed, the *Priscilla* drove her bow into the dock at Fort Michie. The captain was a veteran of these waters and evidently did everything as usual that night; the only explanation for his ship fetching up on Great Gull, investigators decided, was that an unpredictably strong tide had set her well to the south of where she should have been.

I think of how I hid from the wind the other day, choosing to tern farm on what I now consider *my* patch of beach near the dock. Such northwest days along the coast produce a tension in me. Blowing, blowing, steadily and hard, the wind seems to tear at my self-possession. Man's boasts about dominion over nature express not confidence at all, but fear. I remember being aware of this during a New England snowstorm a winter or two ago. The snow began at night. Fine, almost mistlike, it was brushed against the house by a light wind—snow so fine and gently driven that one had to lean close to the windows to hear it strike the glass; walking through it was like walking through a veil. Yet it had an insistent quality; almost immediately it began to form drifts around the house. By morning the wind was blowing a gale, and the fine snow coursed along in streaks and clouds and eddies—ten billion crystals a minute flying past the frame of one window. Beautiful, as that northwest day was beautiful. But there was an endlessness to it. Squire Fred Chase once described a snowstorm as "tedious," and that catches the spirit exactly. It could not be stopped; the storm would have to blow itself out. There's the ultimate threat to man—nature essentially untamable, possessor of a bottomless satchel filled with fogs, gales, hurricanes, blizzards, droughts, floods, disastrous tides, unjudgable currents, earthquakes, plagues of insects, epidemics.

Little Gull Light's foghorn toots softly, and the sounding box hums. The lighthouse also has a radio beacon, now, which helps the big ships that sail up and down the Sound, freighters and oil tankers. Gradually, the sea has been made a safer highway by technology—electricity, steam engines, gas engines, diesels, sonar, radar, radio, weather satellites. But collisions and wrecks still occur. I'm sure that in some ways all this equipment actually has severed connections to the sea—perhaps not for the professionals, the masters of those freighters and tankers, but certainly for the weekend sailor with his unsinkable fiberglass put-put.

That, it seems to me, embodies one of the tragic contradictions in our relation to our only planet. We are frightened by the power of nature, discomfited by the weather, confused by the unpredictable in our environment; we want to live more comfortably, to enjoy life more and enjoy it longer, to travel faster and more safely. So we invent and construct things that have the effect of putting bulwarks and distances between ourselves and the earth. This gives us a feeling of safety, of mastery over nature, and that feeling tends to separate us spiritually and psychologically from our roots. As we lose contact with the earth, we become careless of it; having become careless, we behave as if we would pay no price for our carelessness—as if nature, being "defeated," had not the wherewithal to jeopardize us. But nature is not the sea or trees or grass or birds or pure air and water or butterflies or whales; nature is a complex set of forces that is never defeated. And if, from our safe vantage point, self-dislocated from the earth, we carelessly insult the earth beyond certain limits, we and our co-planetarians will be overwhelmed by those forces.

Our separation from the earth is communal, but our reasons are necessarily personal and selfish. And there lies another contradiction. I am interested, for example, in the strange way some of the human community reacted to the idea of lighthouses. There's a saying (no one seems to know the original source) that "nothing indicates the liberality, prosperity, or intelligence of a nation more clearly than the facilities which it affords for the safe approach of the mariner to its shores." Quite so. Sailing nations depended and depend on trade for their sustenance, so there is a good deal of self-interest involved, but the sea is a highway for the world, and the mariner approaching the shore may come from Sweden or Egypt or Australia or Japan; the benefits of providing for a safe approach accrue to everyone. Only two hundred years ago, lighthouses to mark the approaches and to warn sailors off bad ground were few and very far between, although the idea of lighthouses was thousands of years old. This must in part have reflected the real distances between nations—the distances, in fact, imposed by the sea. It must also have often represented fears of surprise attacks by an enemy navy; sometimes the reefs that guarded your harbor were among your best defenses. But what about the coastal trade, the local fishing fleets, the navy of one's own nation? Why not build beacons and set out buoys for the sake of *their* safety? Well, one reason was cost, and another was profit. Many were the fishing towns where incomes were supplemented with the beached salvage from shipwrecks. Some places, "wrecking" was a trade that had all the moral splendor of piracy: if the coast were a dangerous one, at night—particularly on a bad night—the wreckers would mount two lanterns on a mule and drive the mule along the beach; to the shipmaster offshore the moving lights looked like another ship, sailing much nearer land than he, so, thinking that a course closer to shore would be safe for him, he would steer that way and come to grief on a shoal. If those on shore did not go to such lengths, did not wreck ships on purpose, then of course the event of shipwreck was an act of God—often an act of God the benefits from which were not to be lost because of such inconveniences as the survival of the master or the crew. The ship *Association* wrecked on the Scilly Isles off Britain's Cornish coast in 1707, and Admiral Sir Cloudseley Shovel swam safely to shore, where he was killed by a woman who wanted the rings he was wearing. Rebecca Harding Davis recorded in *Lippincott's* magazine the boast of a Barnegat, New Jersey, wrecker in the 1870s: "No man or woman was ever robbed on this beach till they was dead. Of course, I don't mean their trunks and sech, but not the body. The Long Islanders cut off the fingers of living people for rings, but the Barnegat men never touched the body til it's dead, no sir!"

And these wrecks, these acts of God, should not of course be prevented by any agency of man, when man could benefit so from the result. When in 1619 a Cornishman, Sir John Killigrew, petitioned the British crown for the right to build a lighthouse, the people living near the proposed site opposed it and made the project, Killigrew complained, unexpectedly "Trobellsom." They "affirm I take away God's Grace from them," he wrote. "Their english meaning is that they now shall receve no more benifitt by shipwrack (for this will prevent yt). They have been so long used to repe by the Callamytye of the Ruin of Shipping, as they clayme it Heredytorye, and hourly complayne on me." Nearly two centuries later, along the outer beach of Cape Cod, the same objections were raised. Ralph Waldo Emerson visited Nausett Light, where "Collins, the keeper, told us he found resistance to the project of building a lighthouse on this coast, as it would injure the wrecking business."

One night, listening to a lighthouse foghorn on the Maine coast, I thought of the sailors out in the fog, navigating by their ears (I've been lost in the dark in a small boat myself), and I reflected that the sea demands the simplest responsibility of men for one another. But I tend to have a romantic view of such things. I get a charge, for example, from Kipling's *The Bell Buoy:*

> I dip and I surge and I swing
>   In the rip of the racing tide,
> By the gates of doom I sing,
>   On the horns of death I ride.
>   A ship-length overside,
> Between the course and the sand,
>   Fretted and bound I bide
>   Peril whereof I cry.
>   Would I change with my brother a league inland?
> (Shoal! 'Ware shoal!) Not I!

Kipling? My God, how old-fashioned can you be?

But have we changed so much? What about the wreckers and vandals on Great Gull? They still come out here each winter and steal whatever looks useful. To tell you the truth, I think that this assumption of dutiful mutual dependence between man and man is one of the spiritual roots of environmentalism, and one reason the environmental movement finds such heavy going in its dealings with private industry and private persons is that the spirit of community is fostered by what people get from community and is undermined by what they are asked to give in return.

Years ago, as a journalist, I covered an Army show at Fort Benning in Georgia. Those were the early days of More Bang for a Buck, of dependence on strategic air forces, submarines, and intercontinental missiles, and the Army felt like the nation's military stepchild. The show it put on was called Project Man; leaders of industry and government and hundreds of press people had been invited to see how important the ground soldier remained in the nuclear age. During the show, which lasted several days, Secretary of the Army Wilbur Brucker addressed the assembled V.I.P.'s. Brucker was a round, earnest little man who had an evidently novel view of patriotism. He took the occasion to plead with the industry executives in his audience not to soak Uncle Sam with their military contracts. It is, Brucker implied, your country, too, that the government protects with the arms it buys from you. Keep your profits reasonable. When the secretary was done and the meeting was breaking up, a man rose from his seat across the aisle from me and shouted to a friend some distance away, "I hope you didn't listen to that crap about giving it to them for free."

It depends on whose ox is being gored, and it doesn't matter how damaging to the rest of us that ox may be. Obeying speed signs, for instance. I get the feeling that many people who drive automobiles—including me, too, on occasion—now believe that if they are willing to pay the higher gasoline prices that come with our energy crisis, then it's up to *them* how fast they drive. That denies a concern about the impact such behavior, cumulatively, has on the community. Each of us is a special case, with special justifications for cheating—just as we were before, when speed limits were based only on safety. Can anyone reasonably claim that this sort of selfishness is any different from the selfishness we have all been charging to the companies that produce the gasoline?

"I am pessimistic about the future," said Rutgers zoologist Bertram J. Murray, Jr., a couple of years ago, "not because the problems we face are technically or economically insurmountable but because they seem humanly insurmountable."

And yet, we have always been capable of prodigies of effort. Common workaday prodigies, like Squire Chase's activities here. Prodigies for the public welfare, like Giles Holt's bravery. Another instance, here at the entrance to the Race: on the far side, to the north, is Race Rock Light, built after the Civil War. The builders took eight years to finish that lighthouse—six years just to set the piers, because the structure had to be founded on an underwater ledge where the tides whipped along. And out there at Little Gull Light, now hidden in fog, where the horn is strumming the strings of the air: to build the seawall, after the hurricane of 1815 placed the Establishment in extreme danger, more than twenty-five thousand tons of stone were sailed out from Connecticut.

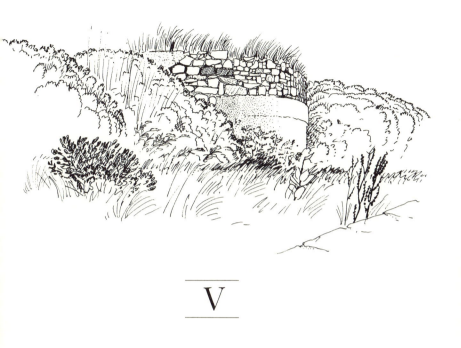

V

May 24: Fog again. The terns in the nesting areas near the dock are much more aggressive than they were a week ago. Many of them have eggs by now, and as we carry up the groceries and the jerricans of water and the backpacks and sleeping bags, the terns dive at us, challenging with their soprano machine-gun cry, *eh-eh-eh-eh-eh-eh*. The people who have been out here through the week meet us at the dock wearing the uniform of daily Chick Check—ancient clothes and battered pith helmets, all well splattered with tern excrement. Everyone looks tired— these are long days for Gull Islanders—but in high spirits. We are twelve this weekend.

May 25: The Great Gull day still begins at first light. I am congenitally an early riser, even after having been wakened in the middle of the night by a thunderstorm, as I was last night. I swing my feet out of the sleeping bag, which is spread on a mattress on the floor of my room, put on socks and white painter's pants and sneakers and a sweater—all while still more or less sitting down, close to the warmth accumulated in the sleeping bag; then get up, stuff my notebook in my pocket, and tiptoe outside.

The thunderstorm presaged a change in weather, but the fog is still with us. I walk down the macadam path toward the Carpenter's Shop, listening for singing land birds, but hearing few, and shove aside the huge wedge of a doorstop that propped the plywood kitchen door shut overnight. There are only a few plexiglass-covered holes in the plywood over the windows in this room, so even with the door open the place is as dark as the inside of a hat this hour of the day. I light a couple of candles on the dining table before turning to the two Coleman stoves that are lined up on a high bench near the door. I get a burner going, fill a kettle with water from a red jerrican under the bench, and put the kettle on the stove. I go back outside to wash my face at a rickety little table where sits a leaky plastic basin and a cracked cake of soap that must have been on the table at least since last fall. The water for these ablutions, poured from another jerrican, comes from a well around the corner; we drink that well water only *in extremis*, when the water brought by the *Sunbeam* runs out, because it is a bit brackish and is thought to be of marginal quality. The drying towel, gray with use, is draped as always over a nearby bush, and it is too wet to serve its intended function this morning, so I reach through an unboarded window at that end of the building and fetch out another towel—drier, but otherwise of similar quality—which hangs there on a string clothesline. Sleep out of the eyes and hair combed, I return to the kitchen to wait, lounging in the doorway, for the water to boil.

Helen Hays appears along the path from the ladies' side of the officers'-quarters dormitory; she waves briefly when she sees me. She is wearing frayed dungaree shorts and a bulky red sweater that hangs below her hips—her usual early-morning costume on Great Gull. "Not much around?" she asks as she reaches the door. Not much, I agree; the thunderstorm and the fog have not brought us many migrants.

The white-bearded Ken Parkes, distinguished curator of birds at the Carnegie Museum in Pittsburgh and a summer regular on Great Gull, joins us for a quick cup of "hot" and then goes down to the blockhouse near the dock to read tern bands. For Ken, too, as for young birders like Tom, Great Gull offers a chance to learn. "Until recently, practically all my ornithological experience and expertise had to do with matters studied chiefly indoors with museum specimens," he says. "My 'field work' was principally collecting trips, obtaining specimens for taxonomic, distributional, and anatomical studies. In 1967, Helen invited me to Great Gull Island for a weekend in late May. I saw both species of terns in their courtship displays, and I learned about how to watch birds *doing* things just on that one weekend. The following year I spent a week in June on the island; I learned how to trap adult terns, I skinned some casualties, and managed to save two out of four downy young spotted sandpipers that had gotten drenched in a storm, by carefully warming them over a Coleman lantern and fluffing their down with a tooth-brush." He was hooked. For the last five or six years he has been on Great Gull for a month or more, all told, each summer, to help in all sorts of studies. "What I have learned on the island, from Helen and from my own observations and from other people here, has *vastly* increased my horizons as an ornithologist."

Now another Helen, Helen Lapham, comes into the kitchen to fetch her telescope, and she heads out into the fog to watch some of her color-banded song sparrows. Ron Franck, a rangy young ornithologist with a pleasant face and a wild halo of curly dark hair, passes down the path, looking for his color-banded yellow throats. Helen Hays goes off, headed for the western end of the island to trap red-winged blackbirds and spotted sand-pipers for banding.

Normally Tom would be up by this time, shaking out the mist nets, but the morning is too foggy; fog wets the birds and, tangled and struggling in the nets, they quickly lose the oil in their plumage, catch cold, and die, so there will be no netting of small migrant land birds this morning until the fog lifts.

I fuss with my camera and carefully clean the lens, hoping that the fog *will* burn off and give us enough light for photography. During the next hour people begin to collect in the Carpenter's Shop. The telescopes Helen Hays and Helen Lapham have been carrying on their rounds now rejoin the squad of scopes on tripods ranked at the back of the room in front of a set of bright blue shelves that bear canned goods. We sit down for breakfast, crowded on benches around the dining table, which is covered by a flowerprint oilcloth and ornamented by a weird wax bird, which wears a defunct Fish and Wildlife Service band on one leg. The wax bird is a candle that has been here for years because no one could bear to light it; instead, it serves as a candle holder and is covered with drippings, particularly on the huge part of its anatomy from which it gets its name, Beaky.

We eat scrambled eggs and fried Spam and doughnuts off paper plates. Hot water is poured into a wild assortment of metal, pottery, and plastic cups, and everyone adds his or her version of instant something. Ron Franck makes a particularly ghastly looking hot drink with a powdered orange-juice substitute made famous by the astronauts. After the meal, cups and pans are stacked in what passes for our sink—a square plastic basin that sits in a sort of counter outside—against one of the cement walls that holds up the earthworks raised to shelter this low building from enemy naval fire.

The morning's work begins again. Ron goes out for yellow-throats and Helen L., for song sparrows. Tom heads for the fire-control room at the sixteen-inch-gun emplacement, to check the equipment for his further experiments concerning the tem-

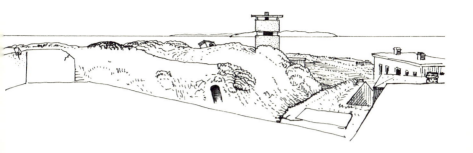

peratures of tern eggs. And now, as the fog at last starts to lift, the rest of us leave for Chick Check.

We go in one party today, beginning at the South Beach, where the most extensive tern farming was done early in the month. Already the wild mustard and beach peas are poking up all over the cleared space, but there are nests everywhere. We step into the nesting area, being exceedingly careful where we put our feet. The first reference in ornithological literature to a common tern colony in Great Britain had to do with a ternery at the mouth of the river Tees, where "an infinite number of sea-fowle laye their eggs heere and there scatteringlie in such sorte that in tyme of breedinge one can hardly sette his foote soe warylye that he spoyle not many of their nestes." South Beach is rather like that by now, and it is particularly unnerving for the likes of me, who hasn't been over this course the last ten days and learned where eggs—many of them the color of the sand and rocks—have already been laid.

The birds at this end of the beach have risen, screaming, at our approach and are diving at us. Ducking slightly, we spread out across the beach, more or less in lanes, with two of the younger and more limber girls in the party deployed among the piled granite blocks of riprapping that separate the upper beach from the water; they will be looking for roseate nests, by crawling between rocks and lying in tortuous poses to peer into caves. And so we advance down the beach, Helen Hays moving back and forth with a clipboard in her hand, taking down the information the members of the crew call to her over the uproar the terns are making.

"Helen—"

She looks up at Ken Parkes.

"New nest. Number 937." From a back pocket Ken has taken a stack of wooden tongue-depressors with numbers already written on them in indelible ink, and he now plunges marker number 937 into the sand, number end up, beside the new nest. "One egg," he says, picking up the egg. He writes the number *1* on it with a marking pen, puts the egg back in the new nest, and moves on.

"Helen—"

Having jotted down the record of nest 937, one egg, she turns to face another Chick Checker.

"Nest number 723. Second egg." The second egg is numbered *2* and put back.

"Helen . . . nest number 454 . . . marking off with two." Enough time has passed since the second egg appeared in nest 454 that it is unlikely that that pair of terns will produce a third egg; for them, the clutch is complete. The number *2* is written on the wooden nest marker.

"Helen . . . 592 . . . third egg, and marking off with three." Helen greets many of these announcements with an enthusiastic "Marvelous," "Good," "That's *wonderful*." The tern farming has borne fruit; this is proving to be a productive season.

"Helen . . . nest 405 . . . marked off with three the other day, but now number three is broken." She crosses to the nest in question, stepping from rock to rock to keep off the cluttered sand; checks the egg to see if the shell is abnormally thin, decides not, asks that it be buried so that no predator, such as a blue jay, will find it and discover the delicious taste of tern eggs. If the egg were not only broken but empty, the culprit might have been a gull or a black-crowned night heron; but it isn't.

One of the parent terns may have done the damage, making a departure from the nest.

As we move along the beach in our ragged foraging line, the terns swirl up before us, unwillingly and protesting, and those whose nests we have passed swirl down behind us. From any point on the island you can judge how far Chick Check has progressed by the terns in the air above our heads. For many of them, we need only move a few steps beyond the nest before they are sitting again, muttering nervously at us, ready to take flight if anyone comes back. I am struck by their guttiness. What a reversal in form has occurred in a few weeks: the same birds that shunned the breeding ground now hate to part from it, even for a minute or two.

They swirl up before us, screaming, raining excrement. Some even strike at the backs of our heads as we crouch over nests. I have been foolish enough to wear only a cloth rain cap on my head, and those sharp bills *hurt*, rapping on my skull through the cap. Terns have been known to kill young gulls and to make life miserable for other birds that way, and now I can feel how.

I am particularly vulnerable because of my assignment. One of the young Chick Checkers, a new birder named Harold Heinz, is serving as my holder-of-eggs. He takes completed clutches from their nests, lines the eggs up in the holes of a graph-paper rig I devised, sets the nest marker beside them, and then holds the rig steady while I set exposure and focus and make the photograph. All that time, both of us have our heads down, so each strike comes as a surprise. I try getting my head at least as low as Harold's, because terns tend to strike at the highest point in their line of fire. But he, too, is wearing cloth headgear—a ratty old Gull Island fedora stained with years of tern droppings. Sharing the wrath of the birds as well as the job, though have met only the previous afternoon, we quickly are fast friends.

Fast friends, but not quick enough to suit Helen. It is a poor idea to linger in one spot in a ternery for very long during the breeding season. Extended disruption now may cause terns to desert nests and eggs. Harold and I are moving with all possible speed, but photographing eggs takes time, and we lag well behind the rest of the Chick Checkers. Frequently we skip considerable stretches of ternery to catch up.

The group finishes the three hundred yards of South Beach and climbs the steep earthworks to "the Top," the roof of the cavern behind the big gun emplacement; common terns are nesting there, at the grassy edges and on patches of gravel and purslane and along the strips of grass that grow between the slabs of cement. Our path is taking us in a rough figure eight— now across the roof, then clockwise around the eastern end of the island, back toward the South Beach along a narrow strip of spottily tern-farmed ground between steep earthworks and a seawall, up over the cavern roof again to North Rocks—more riprapping—then down North Beach to the dock, to *my* patch of tern farm, where, wonder of wonders, there are terns nesting. At the dock we break for lunch, before making the counterclockwise loop around the west end of the island. This is exhausting. Three hours of leaping from rock to rock, being endlessly cautious about where the feet are placed, scrambling over huge granite blocks, balancing precariously to make a photograph or number an egg, ducking the attacks of terns. But it is also fascinating. I have noticed again and again how much more I see when I try to *record* what I am looking at. Usually I do that with a notebook and pencil, but using the camera has the same effect on me. Every few feet, it seems, Harold squats on his haunches and sets up a new clutch of eggs on the graph-paper rig, beside a pulled-out nest marker, and I bend over it, peer through the viewfinder, saying "Fantastic. . . . Gorgeous. . . . *God*, that's beautiful." I sound as exhilarated as Helen does.

The eggs are so *various*—brown eggs with heavy brown splotches, tan eggs with gray and black speckles, gray eggs with brown and black markings, green eggs spotted with brown and black, one that is pale robin's-egg blue and almost unmarked. They may be quite round, or quite pointed, or something in between. And variations show up not only from nest to nest, but even in the same clutch. Because the eggs are so various, it is often difficult to tell a roseate's egg for sure from a common's, unless you know what kind of bird just left the nest, or if you find the egg resting someplace a common simply wouldn't use for a nest. In the main, the roseate's egg is longer and more pointed than the common's, but that isn't a sure test. "It is hence extremely difficult," said George Mackay in 1895, "to absolutely identify all of them." In fact, he was under the impression that roseate and common terns sometimes laid eggs in each others' nests.

Looking at clutch after clutch, I can now see the attraction of the old-time oology. A great many amateur oologists used to fill whole cabinets with eggs, to have them and trade with other oologists, as if these beautiful things were stamps or bubblegum cards or matchbooks.

After lunch: As Harold and I search for completed clutches around the edges of a cement floor—all that remains of one of the Army's many buildings in the long, narrow meadow of the west end—I see a pink egg lying beside a green one. Astonishing. Then I crouch down, pick it up, and discover I am holding in the palm of my hand an egg completely lacking in shell. I've heard about this, even written about it, but I've never seen an example before. The outer skin containing the yolk and albumen is faintly leathery; rolling in my open palm, it feels spongy and exceedingly fragile, and the yolk inside casts a dark shadow on the pink. This is the first of these found on the island this summer—although some obviously thin-shelled specimens have been picked up already. Its chances of surviving the weight of its brooding parents are nonexistent, so I carry it back to the

Carpenter's Shop and set it in a bowl on Helen's desk. It will be preserved, and ultimately it will be sent to a laboratory for chemical analysis.

Midafternoon: Reading bands from the hooded darkness of the sixteen-inch gun's fire-control room. In a driving rain shower, the brooding terns on the cavern roof cloak their eggs with drooped wings, sit with heads tilted up, and now and then shake their heads—a businesslike, not irritated, gesture.

Between showers, there is much activity on the other side of the burlap screen. Terns courting, copulating, brooding eggs, squabbling over territory, flying to and from fishing, begging for fish, trying to steal fish. Roseates pass overhead—I can't see them, only hear them—calling *chivit, chivit* softly. Common tern pairs on the roof are holding what sounds to me like prolonged heart-to-heart conversations in low, growling voices: *kor-kor-kor-kor-kor-grr-grr-grr-kor-kor-kor.* . . .

It is hard to make head or tail of it; but again it conveys to me a profound sense of order—everything in its place. And part of this sense of order, perhaps strangely, is the awareness that the present success of the domestic activity before me is a miracle hanging by a thread. If we have a bad three-day storm any time between now and the fledging of the young—a full blown northeaster—or if the baitfish that are now so plentiful nearby should suddenly disappear, then eggs will be lost, chicks will be lost, nests will be abandoned, possibly abandoned by the hundreds, the thousands. And still these rituals, these gestures and displays, these couplings, these muttered dialogues and screaming fights over territory will continue—if not this year, then next, for as long as places remain to serve as their arenas.

May 26: A marsh hawk drifts low over the island, and the terns climb to meet it, in a rage. The slender, long-winged hawk is no danger now, except to a sick or dying adult, but it will *be* an important enemy when the chicks are hatched. The terns attack it on principle, as they do all birds of prey, as they do herons and blue jays and crows and sometimes gulls. The black-crowned night herons and the gulls are around the margins of the island all day, occasionally supping on tern eggs, perhaps, and watching for the appearance of chicks. Well, predators are preyed upon, too. Ron Franck comes into the Carpenter's Shop to say that he has just watched four great black-backed gulls attack a night heron. They drove it to the water, where it landed—a most unusual roost for a heron—and one of the gulls put down beside it, to ride guard. Evidently the gull was wary of the heron's bill once it was floating on the water, but whenever the heron tried to fly up, the guarding gull attacked it. Ron did not see the denouement. He had tasks to perform elsewhere, and the scene was still being played out after some while. Life does not necessarily make good theater.

June 1: The terns screaming and diving at us as we came up from the dock yesterday were a *greeting*.

I still cannot say, despite my watching and my reading, that I find these complex creatures a great deal more accessible than I did a month ago. There is about them something very much *other*, different, dimension-breaking, not susceptible to captivity in the terminology of human existence. Perhaps that would not seem so if I were to spend more time at it. Ralph Palmer, studying terns in Maine, apparently broke through somewhat; so did a delightfully literate couple, George and Anne Marples, in England in the 1930s; and here, Helen Hays determinedly burrows through for facts, and warns the likes of me against such "blinding" interpretive descriptions as "amorous" and "brave" and "contented"—indeed, against making interpretations at all.

Well, she need not worry. When, off the island, I close my eyes intentionally to seek the terns with my mind's eye, I do not see postures and actions that mean this or that to me. I see light little bodies and sharp wings; they swoop down at me, rebound from the bottom of their dives and rise to a point of no motion, wings quiet, hold there a beat or two longer than my expectations of gravity have prepared me for, then swoop down and rebound again—like yo-yos, each like an independent toy of the sky on the end of a string. And if these strings are still invisible to me—even though whatever truly moves the birds may be invisible to *everyone* now—still, I feel familiar in some way, as if I'd crossed a threshold.

The terns' fecundity this year is spectacular, at least in its timing. Nearly two thousand active nests have already been found, nests with eggs. The terns are a week or more ahead of schedule, based on the experience of recent years, and maybe much more than that if one goes back to the 1820s and '30s. Keeper Frederick Chase used to collect what he recorded in his journal as the Primo Larus Ova at the very end of May or during the first week in June, and not very many of them then, tak-

ing 104 on June 3, 1827; 10 on June 1, 1828; 18 on June 5, 1829; 5 on May 31, 1830; 12 on May 30, 1831; and 8 on June 6, 1832. This June 1 there are thousands of them. The weather has been good, the baitfish are plentiful—sand launces and butterfish, mostly—and the terns have made use of a great deal of the nesting territory we opened up for them.

They hurry to get through to their harvest; the season is short in the best of years. The season used to affect most of our species this way not many years ago, depending as we did on raising our own food. For us as for the terns, the warm months were privately crucial, and bustling. Now fewer and fewer men, using bigger and bigger machines and increasing amounts of chemicals, supply food for more and more people: another force thrusting man away from the reality of the planet—particularly in those nations that have the greatest impact on the planet.

We "progress." At least we change, rapidly, our way of living, by changing *things*. In the process we even lose track of where we've been. Paper crumbles; ink fades; recollections of experience are not written down, and are lost; records are consumed in fire and flood, or are simply forgotten (two federal department heads cannot even get straight the basic history of a piece of land they have decided to build a fort on, even though the basic history survives). Ourselves and our works dissolve, vanish, like Squire Chase's barn on Great Gull, like the house he built here. Where are the paving stones he carted to lay as the floor of the house? Do they now sit in the dry walls that foot some of the Army's older earthworks? Or are they simply buried with piles of other rubble? Where lie the bones of Richard Thomas and boatswain John English of the *Superb?* Were they dug up by the workmen who transformed the keepers' mouse-ridden garden into a seventeen-acre brick-and-cement battleship? Did these discoverers, unearthing bones and skulls, sport with them (the bolder ones chasing the others as if to force the remains on them), then finally hurl their discoveries into the Sound? Or were the bones unearthed by one of the credulous treasure hunters? Much is lost forever, our progress through the centuries trailing out behind like something badly woven, its tail always unraveling. What might we know if it didn't?

The terns may be arriving earlier, nesting earlier, laying earlier. We don't know, although what comes down to us from the last century seems to hint at that a little. And perhaps the roseate terns are only now discovering the comfort and security of nests in rock caves, or the occasional acceptability of common terns as mates. We don't know, have no way of knowing.

Change is part of the planet's continuum, in any case. But for the terns—left to themselves—the changes appear geared to a far slower clock than ours. Killed, robbed, forced off for years at a time, they have persevered in their use of Great Gull while the island went through at least three drastic revolutions in its human use—to farm, to fort, to research station. And this year they are here in force again, and they have laid thousands of eggs by the first of June.

VI

June 27 (three days past Midsummer's Day, when Samuel Wyllys' one fat lamb was due): From the *Sunbeam*'s flying bridge, as
we cross to Great Gull, I see a common tern flip past in front of
us, and it is wearing a red wingtag—a plastic ribbon fastened
around the wing just above the crook or wrist. "Oh, yes," says
Captain John, "that reminds me. I've got a whole list of them
here." He points to the wall of the pilothouse above the wind-
screen, where he has fixed a couple of sheets of plexiglass as a
kind of blackboard for keeping a running record of the fish
caught by his customers; today one column of his crayoned hen-
tracks, undecipherable by me, refers to terns that have been
wingtagged on Great Gull and then been spotted by Captain
John and other fishing-party-boat captains on Long Island and
Block Island sounds. A young Harvard student at Gull Island is

studying the fishing habits of the common terns as they feed their young, and since the island has no work boat of its own, the researcher has to hitch rides on other people's boats when he can, in order to track the wingtags, and also have others do tracking for him.

Captain John switches his citizens-band radio to our channel, picks up the microphone, whistles into it, and says, "Hello, Gull Island. Anybody home?" The station has its radio on, in expectation of the *Sunbeam*'s arrival, and Ron Franck answers. "Go ahead, Captain John." "I've got some wingtags for you. . . . June 22, 0830 hours, one flying east just north of Valiant Rock Buoy. . . . June 23, 1305 hours, three fishing, with six other terns, east of Little Gull Light. . . ." He finishes the list, gives Ron our expected arrival time, and signs off. This has become a regular event, and the tracking project simply couldn't be done without Captain John's help. All the fishing-boat captains think there is something ridiculous about these bird watchers on Great Gull; even John has his doubts. But he is an accommodating guy, and since he makes many of these reports by radio and doesn't joke about it, a few of his colleagues have got the drift and decided to feed him data for his reports. As a quid pro quo, the Gull Islanders have taken to calling John on the radio whenever great crowds of terns are fishing within sight of Great Gull, thus letting him know where there might be some action in bluefish and striped bass.

John Wadsworth is descended, it seems, from a gentleman who served on the Connecticut General Court with Samuel Wyllys. A nice closing of a circle. He's a stocky, breezy, wisecracking entrepreneur, with small businesses going afloat and ashore. He has a beard like Solzhenitsyn's, usually wears a red nylon windbreaker with "Captain John's Sport Fishing Center" printed where the breast pocket ought to be, and covers his bald crown with a long-billed fishing cap. During the trips out and back, I join him on the bridge, catch up on how the fishing has been, and listen to him carry on at least two conversations at once with his buddies over his radios.

June 28: There are twenty-eight hundred tern chicks on the island this morning, and more pipping every hour. It takes from eight to thirty-six hours for a chick to open its shell. Then the egg breaks; the wet, peeping little bundle of down falls out into the open. Usually within a few minutes, the brooding parent or even a neighbor picks up the broken pieces of shell, flies with them out over the water, and drops them. This has the effect of removing evidence that might catch a predator's eye; it also eliminates a magnet for ants (ants can make a chick's life misera-ble . . . and short); and it removes a danger to unhatched chicks: a broken shell half, its inside wet and sticky, may roll and get jammed on another egg and thus present the chick inside with a well-glued double thickness of shell to break when its time comes to pip.

So the chick, smaller than a ping-pong ball, emerges, crying for food and attention. Its eyes are nearly shut, feathers wet, legs too weak to carry it. It is brooded under the adults on the nest and protected from the weather and the sight of predators. Its feathers dry; it becomes fluffy brown with black markings. It struggles with the food its parents bring; often the baitfish is just too big to swallow, and the parent, after trying a few times to get the chick to take a fat three-inch butterfish or the like, drops the food and goes off after something better. Often the food will be narrow enough to slip right down the little red lane, but will be too long, and the chick will sit, looking foolish, with the tail of the fish sticking out of its mouth, and will get the whole thing inside only after the fish head has been digested to make room. Fed at least once an hour when the fish are plentiful, the chick will grow from a tiny item an inch and a half long and fifteen to eighteen grams in weight to a monster—known here as an "Ele-phant"—seven inches long and more than one hundred and twenty grams in weight in the space of twenty days. About twenty-five days from hatching it will have lost its downy feathers and pushed out the feathers of an unfinished-looking gray and white plumage; it is now just able to fly, and Gull Islanders call such terns "Orvilles."

Told like that, the process sounds charmingly simple and domestic. But in fact these twenty-five days are days of hairline existence for the chicks. Many of them die. Some of their parents are not efficient fishermen; perhaps a three-day storm sweeps in and makes the fishing difficult or impossible; and if both parents are off the nest looking for food, the storm may finish the chicks before hunger does. After the first day of life they are strong enough to skedaddle into the nearest clump of grass or under sheltering rock or bayberry bush, to hide from the rain, the sun, the cruising herring gull and black-crowned night heron; but they may not be safe even so. The chilling rain and wind may be just too severe and go on too long. The heat of the day may make the whole beach an oven, shade or no shade. Oliver Austin, Jr., wrote of an experience he had before he knew how vulnerable tern chicks are to heat: "I tethered one five-day-old chick under a trap in the open beneath a broiling sun in a well-meant effort to catch its parents. The old birds were probably both off fishing, and under normal conditions the chick would have retired to the shade of a clump of grass, but here, tied in the open, it lasted just twenty minutes, at the end of which time it suddenly stiffened out and died. I did not repeat the experiment." And the parents are helpless to scare off any of the larger predators, once the big birds have spotted a chick they want. The whole colony may dive on the feeding heron, screaming imprecations and banging on the enemy's head, but the heron is unmoved, and finishes its meal. If they are so powerless in daylight, the adults are worse than helpless against a nighttime predator. Ian Nisbet of the Massachusetts Audubon Society recently told me about a colony of terns in

Massachusetts that one summer was raided each night by a great horned owl. After the first couple of nights, the terrified adults in the colony all left the breeding grounds every night, despite the fact that they had chicks and eggs in the nest.

As the chicks get old enough to wander, they explore beyond the boundaries of their parents' nesting territory. When such peregrinations take a chick into the territory of an adult tern that is very protective of its boundaries, the chick may find itself under a determined attack, being pecked hard by the adult, and then have to run a gauntlet of irritated, pecking adults on its uncertain, terrified race for home; it may be killed before it gets there.

If it is not well, or not aggressive enough, it will quickly become too weak to survive; it must, in order to be fed, attract its parents' attention with its voice, and peck determinedly at the bill of a parent bearing fish. And if it does become too weak to survive, its parents may kill it themselves before it starves, by pecking it.

Even the chick that survives the twenty-five days and reaches Orvillehood may still be in grave danger. "I have seen juvenals . . . circle out over the water and have to alight there for lack of strength to get back to land," wrote Ralph Palmer. "When a strong tide was drawing out past the islands, such birds were carried away to sea, their plumage became soaked, and they drowned. When they joined an up flight [a general-alarm take-off] in a strong breeze which they could not fly against, they suffered the same fate."

I once was watching a group of just-fledged and adult Arctic terns as they bathed together near Matinicus Rock in the Gulf of Maine. Into my binoculars' field of view flew a great black-backed gull; the gull swooped into the swimming flock, and suddenly all the terns in the flock were up and chasing the gull—all, that is, but one of them, an Orville that hung, broken, from the beak of the gull.

Chick Check, when there are new chicks as well as new nests to worry about, is even more alarming to the neophyte researcher than when the nesting territories are thick only with eggs. It is one thing to step on an egg and break it, or to drop it after you have picked it up to write a number on it; it *is* warm, it *is* alive, and you feel badly about breaking it, but the yellow blob of life inside hasn't taken form yet. Stepping on a live chick, or injuring one in the process of banding it, is quite another matter. When we move among the nests, with parent terns screaming and dive-bombing us from above and the chicks toddling off in all directions toward what they hope will be good cover, every step has to be taken with great care. We move slowly, scrutinizing the ground, and set our feet down gently, in the foolish hope that if our eyes have deceived us we will not hurt the chick beneath our foot. A few chicks *are* killed each season because of Chick Check. For one thing, those of us who are unpracticed are unsettled by the constantly shifting hazards in front of us, lose our concentration, and move hurriedly. Helen has more than once called me to cross a beach and make a photograph, and then, as I started toward her immediately with my eyes on her, shouted, "Don't look at *me*, don't look at *me*, look where you're *going!*" (Nicely symbolic, come to think of it, of our relation to the planet.) Or, chasing an unbanded chick that has scuttled into the long grass back of its nest, one may miss seeing the other chick that has already hidden itself in the way.

We are numbering new nests today—now up almost to three thousand of them on Great Gull, including some that were abandoned when the parents made a second start for one reason or another—and we are banding chicks with the single, numbered plastic bands they will wear until they are Elephants, when their legs will be big enough to hold the four-band combinations. Because chicks are too weak to travel out of the nest in the first twenty-four hours of life, the daily Chick Check allows us in most cases to know which chick came from which egg in which nest; and after an individual Elephant gets its adult bracelets, having been put through each of these steps, there is then a written record that traces, say, B/A × RB/Y (Blue over Aluminum × Red-Blue stripe over Yellow) right back to the nest it was born in and the egg it hatched from. The record often shows if it was laid first, second, or third in sequence; we know the kind of territory its parents chose for a nest site. Probably at least one of its parents has been followed in the same detail, and we may be able to discover the parent's identity. And when B/A × RB/Y returns to Gull Island in a year or two or three, researchers can then observe its behavior in the light of what is already known about it. If this process continues day after day, season after season, tern generation after tern generation, with thousands upon thousands of nests and eggs and chicks and Elephants and Orvilles and breeding birds, the accumulated data can become extremely meaningful and, in a way, creative all by itself, because of the patterns it suggests. Helen points out matter-of-factly that one great virtue of this study is its longevity and persistence. A lot of this sort of work elsewhere is done by graduate students pursuing Ph.D.s; their studies are made over the space of a few years, the degree is either achieved or not, and the project ends. The doctoral dissertation or other paper that comes out of the project may be very useful, but in many cases, if the young scientist is bright and perceptive and imaginative in his approach, his work raises a great many new questions, perhaps even more than it answers of the old questions. And these new questions often have to wait for answers until another doctoral candidate decides they are worth pursuing. The Gull Island research project has been under way for ten years; it has been carried out with its present intensity for five years. Helen apparently has no idea when the data will be sufficient; or if she does, she isn't saying. Quite likely, the results of the first ten years of study will suggest avenues that will take at least another ten years to follow, and those ten years, ten more. The "most enticing feature of the work," wrote Oliver L.

Austin the elder in 1934, "has been the repeated refutation of concepts and deductions by subsequent observations." And the Reverend Gilbert White, who studied and theorized about wildlife in Selborne, England, two centuries ago, commented in 1770: "Though there is endless room for observation in the field of nature, which is boundless, yet investigation (where a man endeavors to be sure of his facts) can make but slow progress; and all that one could collect in many years would go into a very narrow compass."

Loudly, over the vociferous protesting of the terns:

"Helen . . . new nest . . . number 2968 . . . *two* eggs."

"Okay. Mark them both *A*."

"Helen . . . nest number 2815 . . . chick has hatched from egg number two . . . getting band number 384."

"Good; 384, right?"

"Right."

"Helen . . . dead chick." Harold Heinz rises from a crouch, studying the desiccated little corpse in his hand; holds it out for Helen to inspect, then puts it in a shirt pocket. It will be preserved for chemical analysis in the lab.

"Helen . . . unbanded chick; don't know which nest."

"Okay. Common?"

"What?" We really have to shout to make ourselves heard; the voices of the frantic terns hovering over us seem to swallow up other sounds.

"Is it a common tern?"

"Yes. Getting band number 385."

"Okay. Good. Let's move on a little faster." The need for thoroughness must be balanced against the welfare of the chicks, particularly on so hot a day as this; we must limit their alarm and keep their food-carrying parents from them only briefly. Soon we are off South Beach and on our way up to the roof of the underground cavern back of the great gunpit.

After Chick Check, the hardier types go for a swim off the dock. Like me, Ron Franck does not join in, complaining that the water is too cold. Several hours later, as most of us sit resting in front of the Carpenter's Shop, he comes past us at a dead run from the direction of the dock and races up into his room at the base of the tallest cement watchtower. A minute and a half later he reappears, barefooted and in blue bathing trunks, and runs past us the other way; he gets a scattering of cheers from the recent swimmers, but he does not respond. In ten minutes he is back again, dripping wet, carrying one of the wingtagged terns. He has, it turns out, just rescued the bird from drowning. Bathing and preening itself a couple of hundred yards off shore, it had managed to get its bill stuck under the staple that locks the tied wingtag in place, and it couldn't pull free. It would not be able to fly, hung up like that, and it would have drowned eventually, but Ron noticed its struggle and swam out to get it. David Duffy, the Harvard undergraduate who runs the wing-tagging project, redoes the tag, tying the plastic ribbon loosely above the wrist of the bird's wing and then restapling it. "This happens now and then," he explains, after the tern has been released, none the worse for the experience. "It's a kind of freak situation, but we're trying to find flatter, smaller staples they can't do that with."

No, there is no gainsaying the fact that wildlife research—even such benign, largely observational work as this project—does present some hazards to the birds, and it does result in some mortality. There are researchers on Gull Island this summer, as elsewhere, who feel that such mortality, while it is very small, should not be mentioned in public, for fear of upsetting, well, the "bird-watchers." Little do our local fishermen know it—to them a bird-watcher is a bird-watcher is a bird-watcher— but there *are* such divisions among ornithologists, particularly between the hobbyists and those who have made birds their life's work, and the divisions stir strong passions. Out here, where working with birds is a job, one accepts the few deaths we cause—not without regret, sometimes intense regret, but as unavoidable in such work. It is part of the price paid for knowledge that will lead, if to nothing else, to better wildlife management practices, benefiting the birds in the long run much more than if we left them alone. Here, it also happens to be part of the price paid for the thriving existence of this tern colony. But there are those who see any man-caused death of birds as indefensible, regardless. And their attacks, combined with the researchers' regrets (Could this or that fatality have been *prevented?* each one asks, knowing that some certainly could), make a number of researchers very uncomfortable, defensive; they would just as soon no one talked about it outside the family.

I raised the question with Dr. Frances Hamerstrom of the University of Wisconsin a few weeks past. Fran Hamerstrom has been banding and color-marking birds in various studies for years. She answered:

> You are right; some birders feel that banding, marking, etc. endangers the birds. They *feel* this so strongly that they would be happy to write to their congressman or anyone else with influence, but they *won't do their homework* and get the *facts.* I find this immoral.
>
> Furthermore, their concern for each individual bird is so strong that they prefer not to recognize that without research some species may disappear from this earth forever. . . .
>
> It has never occurred to me not to air these questions. On the other hand, I would be as scared of airing these questions in front of an emotionally-charged anti-research group as I would be of giving a talk on wine-tasting to the Women's Christian Temperance Union.

Amen. One of life's little piquancies for the scientist.

In the late afternoon, song sparrows and yellow-throats lisp furtively in the bayberry. The vegetation on Great Gull is luxuriantly, rampantly green and blooming—purple nightshade, white and yellow melilot, red clover, even the few day lilies someone brought out and planted here last summer.

Feeling tired and well occupied, I carry my telescope down to the dock house to read a few bands through its windows and watch the end-of-day activities, human as well as avian. All over the island in these last hours of sunlight, there is a big push on to trap and band and wingtag adults that are unmarked as yet. One mist net is regularly kept open along the shore during the day, to catch shorebirds and terns, but the terns learn very quickly where that net is, and in order to band and tag a lot of adults, Gull Islanders set out wire mesh traps over nests that contain eggs and young. These are simple enough traps, with down-sliding doors tripped (if they are working properly) by the parent bird walking through the opening and stepping on a treadle. Several of the traps are set out now in the patch of nesting territory just west of the dock house, and I watch the operation between scans for band combinations. Here comes one parent, returning from successful fishing. It lands on the wrong side of the trap and paces around looking for the route to the chicks; finally it finds the doorway, walks in, the door slides down behind, a chick is fed the fish, and the adult tern settles into the scrape, unconcerned, to brood her young. The bird does not become alarmed, in fact, until the young volunteer who is trap setting and checking on this side of the island comes back; then the tern tries to take off, and finds itself penned in.

Not far away, another parent, behaving as if it didn't like the looks of things around its nest, steps through the opening rather tentatively, trips the door before getting well inside, and manages to back out. Now it walks nervously around the closed trap, trying to find a way to reach the chicks. A second adult flies in, and the two of them strut anxiously near the nest, fly up over the trap to study the situation, then land and pace some more. Soon the trap checker returns to set their lives to rights, and collects the birds that have been caught. She will take the trapped birds back to the Carpenter's Shop, where Dave Duffy is doing a land-office business in wingtags this evening. . . .

The traps near the dockhouse lie tipped over, through for the day. Before me, the terns go about their business, feeding young, preening, resting, flying out to the Sound and flying in with fish, brooding eggs and young, posturing near nests, muttering, screaming, chasing, courting. I have been amazed by what these birds will put up with for nesting sites, considering the many years after the war that it took them to decide the island might be acceptable again. A few feet away from me, for instance, below the slit windows of the dock house, the cement platform that once must have been connected to the foot of the long dock is broken, apparently because one corner of it was undermined by a high tide; in the crack made by that break, a common tern has chosen to try to nest. Other birds are nesting around the edges of a cement floor between here and the two officers' quarters, and they do that all over Great Gull. There's even a nest in the rubble of the tipped-over blockhouse out on the west end. I wouldn't be a bit surprised if eventually nests were found on the flat roof of one or another building.

Beyond the platform, the beach—what's left of it—is a clutter of stones and bricks and broken cement and driftwood and ancient lumber and rusty bolts, and there are terns everywhere in that little patch. It's a busily domestic place. A chick stumbles around its nest, unsteady on the rocky ground; it reaches its standing parent and waddles underneath and out the other side. Another buries its head in the side of a parental breast. Two more nestle into the shelter of an adult's hooded wings. As the gray and white parent birds fly lightly, easily, in from the sea, carrying slivers of silver crosswise in their bills, the chicks crouch and beg for food, peeping and shrilling. When the adults land, they are mobbed by their open-mouthed offspring: *Me, me, me, me.*

July 12–14: It is now possible to see the whole breeding cycle from a single vantage point. Whichever patch of beach one looks at, there are stubby-looking Orvilles just able to fly standing beside adults still courting. Chick Check is much shorter; we spend most of it chasing Elephants, groping for them under bayberry bushes and in crevices in the rocks and along tunnels they have made in the beach grass, and those we catch get their four-band combinations. On the move, legging along like miniature ostriches, Elephants are as quick on their feet as we are; and once they are hidden, they can become almost invisible, so that one finds them only by touch and by chance.

Meanwhile, we continue a daylight-hours watch of three tern nests—two roseate and one common—that sit side by side at the top of a tall dry wall built into one of the Army's earthworks. The watch has been under way now for nearly a month. The members of our varying volunteer crew work in two-hour shifts, climbing up into the blind at the top of a tall wooden tower and peering out at the nests eighty feet away. Helen is chiefly interested in which of the parents does most of the feeding and how often food is brought and how big the fishlings are; if we get a good look, she would also like to know what kind of fishlings are being caught, but the terns do not give us a lot of time for that. Through the slits in the burlap over the blind's windows we see perhaps the last two or three seconds of the parent birds' approach; the chicks are already scrambling out of their hiding places in the grass back of the wall, begging frantically, and the chick that gets the food swallows it almost as quickly as a person can blink.

These full-time watches over particular nests on Gull Island have been conducted year after year. This one began when the chicks were hatched, but sometimes nest watches have been started as soon as the first egg was laid and have continued until the last young bird was sufficiently on its own to stop hanging around the nest site. Dozens of notebooks have been filled with consecutive observations, creating what is surely one of the most complete records of nest life in the annals of ornithology.

After supper each evening Helen lines up the schedule for the next day, asking for volunteers. Nest watch begins at first light and ends at last light—fifteen or sixteen hours a day, eight shifts. The observer perches on a stool in the tower blind with a pencil and a small notebook in easy reach, binoculars in hand, a drugstore wind-up alarm clock lying face up on the floor. The time of each arrival of a parent tern is recorded, along with everything else we can see about the feeding. Other events are noted occasionally, depending on the experience and temperament of the watcher—and not just those involving the terns but also data on any red-winged blackbirds and song sparrows that may appear in the view.

The two-hour shift goes very fast for me, even when there are long pauses between feedings; a lot is happening if you look for it. I recently watched fascinated as two roseate adults, which normally perch on the wall ten or fifteen feet apart in front of their neighboring nests, squabbled over boundaries. One might have thought that they had worked this out a month or more ago, but a constant reinforcement and pushing at the edges of territory is the rule, not the exception. What struck me about the encounter was that the *air space* above the nests is evidently included in territory. Once the argument started, one of the ro-

seates spent several minutes in the air, trying to find an approach to its nest that the neighbor would not object to. And then, just as if nothing had happened, they were back in their accustomed places on the wall, facing into the breeze and waiting for their respective mates to return from fishing.

I'm afraid that my interest in such matters has cost Helen a certain amount of the data she is after. Now and then I have looked up from describing nonfeeding behavior in the notebook just in time to see an unidentified parent tern, who wasn't there ten seconds ago, leaving the wall while an apparently just-fed youngster hustled off into the grass again. It is something of an embarrassment to have to report that missed opportunity in the notebook and just record time, nest, and the fact of a probable feeding, immediately below my intense description of something less wanted.

It's too bad that this research has to be done by human beings.

In the twilight I join Harold Heinz as he checks the barn swallow nests in the tunnels and bunkers of Great Gull. Harold—a slender, reflective, private young man—provides a classic example of one important side effect of the work out here. He was just recently introduced to birding by a student friend who happened to be a Gull Island veteran, and he was immediately interested in birds, so much so that he decided not to return to college for his junior year but instead to take a leave of absence, do field work, and try to decide whether ornithology should be his career. As the first step in this experiment, he was to accompany his friend on a tour of the shores and islands of Long Island Sound, looking for osprey nests, part of a censusing project sponsored by the Massachusetts Audubon Society. While you're waiting, the friend suggested, why not spend a little time on Great Gull? So he did. He came out one weekend in May, and he's still here, although he did leave for a while to do the osprey nest survey. He likes to work on his own when he can, so Helen asked him to finish the year's study of barn swallow nesting and young. Harold did with his barn swallows essentially what is done with the terns: the nests were numbered, the eggs in each were numbered, the young swallows were banded, and each evening Harold has been working his nest route, removing the nestlings for a few minutes to weigh them with a set of small scales, and keeping daily records of the whole process. When he's done, he will have weight charts for a slew of young barn swallows. Right now, it appears that the swallow chick becomes an Elephant, too, being fed so much that it weighs more than either of its parents. Then, two or three days before it flies for the first time, the weight begins to drop; the loss probably reflects intentional starvation by the parent birds, which encourages the young to fly; but a change in metabolism may also be involved. In any case, questions, perhaps, for someone to study. But aside from any practical contributions this work may make, Gull Island has once more given a potential young scientist an opportunity to test himself and explore the possibilities before him.

I am reminded of the quiet, sunny-faced Grace Donaldson Cormons, who climbed ashore from the *Sunbeam* with Helen and me on April 29. Like me and many others, Grace can make only occasional visits to the station—a weekend or a week at a time usually—because she has other work to do, in her case a job in the education department of the American Museum. She was, to all intents and purposes, led into serious work in natural science by her first visit here eight years ago; she had been interested in the subject before then, but she was quite young, and she got her direction from the experience. She has published articles in ornithological journals about various observations on Great Gull, and she has continued coming here in her free time. She's now married, to another naturalist, and she is pregnant with her first child, but she was here at the end of April to help start the summer's work, and she and her husband have since been out and back at least twice.

Just now, there are others of the same ilk here. Ron Franck, whose study of that black-masked little warbler, the yellow-throat, has this summer affirmed something that had been suspected for thirty years: yellow throat males are sometimes bigamists; Dave Duffy, in his fifth Great Gull summer, and now mapping the routes and ranges of feeding terns; Tom van't Hof, with his banding, his egg-temperature studies, and his painting, though he has had very little time for painting since May. And a Radcliffe junior named Kathy Duffin, who skipped this Friday's supper to finish the skinning and preservation of a common tern that had been found dead; she sat at a table in the banding-room-*cum*-office, with the Coleman lantern hissing and blazing, and cut and sutured and powdered the skin, all with the fiercest of concentration.

There are, of course, more Gull Islanders than the young scientists. Some are professional ornithologists, like Ken Parkes. Then there are the amateurs—most of them young, but a few, like myself, who are approaching fogeyness in body if not in spirit. Some of us are not really "birders" at all, or used not to be, anyway; among those are two young women who have been working as volunteers in the museum office of the Great Gull Island Project, where they helped organize the mound of data from the island. It seemed a good idea, Helen said, for them to

see where all that data originated and how it was collected, so they visited this spring, and they have kept coming back.

The place is like that. The accommodations are rude, to say the least; the diet is limited by the absence of refrigeration; the hours are long. But the feeling of community and purpose is palpable. Perhaps this is partly due to the youthfulness of most of the volunteers; even the scientists among them have not cultivated those desperate needs to compete for recognition and perquisites that established professionals sometimes exhibit at other research stations, much to the disruption of good-fellowship. In any event, other forces lead. There is first that sense of order the birds impose; and there is also a mutual affection among committed naturalists whose lives center around a routine of observation, learning, chores, roughing it, and sharing.

The island is dotted with observation blinds. Most of them consist of metal scaffolding covered with white canvas, temporary hiding places big enough for one or two observers, and they are likely to be moved a dozen times or more during a season. But on the top of the earthworks back of the Carpenter's Shop, Dave Duffy has built this summer a more elaborate structure, a blind on stilts that overlooks the South Beach, so he can watch the comings and goings of wingtagged birds there. The blind leans down the slope a bit, toward the beach, making the occupants of the platform grateful for the guy lines staked in behind it.

Harold and I were each awake early on Sunday morning, and we climbed up into Dave's blind just after sunrise. The day was hazy and warm. Off in the Race, beyond Little Gull Light, there were already forty-eight boats, big and small, bearing fishermen doing their best to kill all the bluefish and striped bass in Long Island and Block Island sounds. Within a few hours, hundreds of craft would be anchored out there. We sat in the blind for some time, spotting the stub-winged Orvilles and the new-fledged barn swallows in flight around us, watching the terns rise and settle with the passing of researchers on early-morning errands below us. Soon, sometime after breakfast, I would be out in that landscape myself, chasing Elephants again. It was nice to idle a while, apart from it, soaking up perspective and being glad I was on Great Gull.

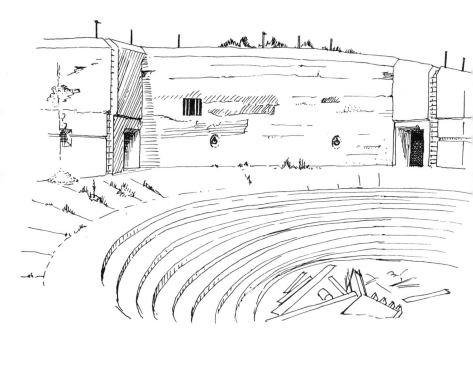

VII

August 2: Relative quiet has settled on the place. Few birds are still brooding eggs and feeding chicks. For many of the terns, the ties to the island have already broken; they have scattered to forage along the coast, and they do not return here in the evenings. Dave Duffy and one of the weekend volunteers left the island with Captain John after we had unloaded the *Sunbeam* this afternoon; John will drop them off on a small island near the Long Island shore, where there is another tern colony, and they will spend the next two days there, making notes on the Gull Island terns that have gone that way.

Doubtless, some of the terns have drifted in the other direction, north to the southern New England shore and the Cape, as they do this time of year. Some have probably begun to move slowly south and are beyond Long Island now. For the rest, the ties loosen. When we came up from the dock, carrying our gear and a week's worth of food, the birds roosting in the patch of nesting territory near the dock house flew up as a cautionary measure, but they were not interested enough in our presence to make more than the most desultory passes at us.

August 3: Harold and I spent some time this afternoon exploring the mysteries of the sixteen-inch-gun emplacement. From above, it is shaped like a square with a part of a big circle at one corner—an arrowhead with swollen glands. The circle was the center of things; that was where the gun stood, swiveling on its track in the amphitheaterlike pit in the middle, which is now filled with green water and gray lumber.

With a ball of string, and the nose-to-fingertip stretch as a yardstick, Harold and I took rough measurements of the emplacement. Along its long axis—the point of the arrowhead to the far side of the circle—it is one hundred and thirty-five feet; the diameter of the circle in which the gun swiveled is about ninety feet. I stood with my back against the wall so that Harold could guess its height; then we changed places, and our best estimate was that the emplacement is sixteen or eighteen feet deep from the top of the wall to the main floor. I also wanted to know how deep the pit was, where the gun stood on its disappearing carriage. The circular steps go down about twelve feet, and from there Harold inched out toward the center of the pool on the jackstraws pile of lumber; using a stone tied to the end of the string as a lead line, he found bottom nineteen feet below the surface.

The history of the sixteen-inch gun rather exemplifies the

whole history of Fort Michie; an elaborate structure, little used. When, in 1945, the two-gun battery above the Carpenter's Shop—Battery Palmer, twelve-inch guns on disappearing carriages—was classified No Longer Required, one of its guns, installed in 1899, had fired only one hundred sixty-nine rounds in its life, and the other, which was installed in 1931, had fired only fifty rounds. They, too, were outmoded in World War II, and their role in the coastal defense of Long Island Sound was given to more modern weapons at Fort Michie.

It is tempting to ridicule the fort as useless, but that's too easy a target. Probably the decision to build a fort here in the first place was taken too quickly, without due consideration of the isolation and bleakness of such an island station, particularly in winter, and the frequent difficulty of landing here. But had an enemy fleet tried to approach New York in 1941, we would have been glad we had such a fort.

Man is a territorial animal, after all. He has ritualized—in

law, within tribal boundaries—the reinforcement of his personal property boundaries, and that has worked rather well. The individual's part of the social contract with the tribe, the "debt he owes his King and God," has been to defend the tribal boundaries. To some extent, tribal territorializing has also become ritualized ("I've got as many missiles as you have"), but because of the nature of the weapons, the ritual has become exceedingly dangerous to the survival of the species, and one can only hope we evolve some less hazardous approach before we blow ourselves up.

There is an atmosphere of humdrum domesticity about Great Gull Island this weekend, of a hiatus in the frantic pace the terns and the other birds follow. The courting and breeding urges diminish; the migratory urge has not fully taken over. It is as if the birds were dawdling, waiting at the station. The Orvilles pester their parents—or, for all we know, all the adult terns that come near them—for the food that by now they seem equipped to fish for themselves. The young barn swallows chitter after their parents, too, and are fed in midair. A great many young birds apparently pass through a long period in which they do not feel independent, however capable of independence they may look to us. I've seen evidence of this with cowbirds and jays and chickadees and gulls and long-legged shorebirds, and probably if I were more alert and spent more time looking for it, I would have seen the behavior in many other species. Young birds have a way of calling attention to themselves that is ludicrously like certain children: they whine and whine and chase their parents hour after hour, insisting that attention be paid and food be furnished. The parents ignore them when they can and sit facing the other way. So Great Gull looks like a neighborhood of families full of adolescents, in which all the adults wish this phase of life would be over pretty soon, but meanwhile there's a routine that everyone has settled into.

However, we can see dramas being played out over the Sound

that once again remind us not to judge these matters in such a superior fashion. Any appearance of a rest period or of comic relief is an illusion. The jaegers have moved in, not many of them, but still in greater numbers than anyone in the Gull Island Project has seen before out here. The jaegers—"jiddy hawks" and "gull-chasers," as they used to be called by the Cape fishermen—are cousins to the terns; they are big, dark, angular birds with hooked bills and hawklike talons and a hefty, powerful look to them. They can outfly almost any tern they have a mind to; a tern passing with a fish in its bill is intercepted, and every rapid twist and roll and dodge the tern makes is matched and bettered, until the harried little bird drops its fish and the jaeger swoops to pick it up. Bullies they may be, these jiddy hawks, but they are the most magnificent ocean predators I have ever seen. Changing direction to stay with their quarry or to catch a dropped fish, they do wingovers and flips and somersaults with such speed and control that one imagines they must have invented the art of flying.

Beautiful to me. A plague to the terns. At no time is life a picnic for any bird—hawk or heron, oriole or sparrow, tern or jaeger. " 'Free as a bird' is an expression in which a bird might find ironical amusement," commented Louis J. Halle, Jr., years ago; "especially as coming from man, the only animal who has, in his individual life, succeeded in achieving some measure of independence from the discipline of nature." I think perhaps Louis Halle's generation rather overestimated the extent of that independence, but it is certainly true that the discipline of nature, for birds, is strict in a multitude of ways unfamiliar now to man. The sun and moon and stars and spinning earth create fields of force in which the birds are trapped; driven and steered by these influences, they move across the face of the planet, they breed and raise their young. And as they voyage, as they court and nest and perpetuate their kind, food *must* be had regularly—for most species far oftener than once a day. The food may not grow well one year, or it may disappear suddenly—

shoved by its own relation to the sun and moon and stars and spinning earth, or by its own need for food—or it may be hidden or even killed by weather. And weather wears birds, exhausts, picks up and hurls: hurricanes deposit Caribbean birds on Long Island and the Cape and Nova Scotia; and migrating land birds by the . . . who knows how many thousands? . . . are blown out over water and never see land again. Predators chase, steal food, kill, and terrify into watchfulness.

More than ten thousand eggs were laid on Great Gull this summer. Perhaps half of them hatched. And in the brief season between then and now, well more than half the chicks that hatched died; only two thousand or so survived to fly. The vulnerability does not end here. The margin of forgiveness is always slight for the weak bird, the less talented fisher, the less lucky, the tern whose inherited store of information lacks something.

There is no way of knowing just what the attrition rate will be between now and the time this year's crop of new terns comes back to Great Gull to nest. Perhaps we will see one or two live to breed eventually of every ten that leave this fall. But that's only a guess. There are too many variables involved. Some birds from this harvest may return north to breed but choose some other colony—on the Sound, on the Cape, in Maine, on the Great Lakes—survive, all right, but not for us. Unfortunately, there are too few of the weekend field birders who pay enough attention to what they are seeing and are conscientious enough to inform the Fish and Wildlife Service's bird-banding office about color-banded terns. A few years ago, Helen spread the word through various birding publications that reports of color-banded terns in late summer and fall would be very welcome. There was little response. Most of what she knows about the dispersal from Great Gull she and Grace and Dave Duffy have learned by chasing the birds in person. Their farthest-north records are from Cape Cod; their farthest-south, from Cape May, at the southern tip of New Jersey. Between Cape May and the wintering grounds, nothing; the terns might just as well have been shipped as freight, unlabeled.

Two weeks from now the terns of New England will begin to head south in earnest, flying in tens and twenties and hundreds, sometimes very high. Gull Island's birds will spend the winter in Central and South America—though, here again, no one knows very much about it. Gull Island band returns have reached the Fish and Wildlife Service from as far away as the Atlantic coast of Brazil and the Pacific coast of Colombia. Most such recoveries are of shot birds—birds killed mainly for food—and seldom does a hunter take the trouble to send in the numbered aluminum bands as the instructions printed on the rings request—if he can read the simple message; so these returns do little more than hint at where the terns go from Great Gull. In fact, all but a few of the recoveries come from one little village, Whym, in Guyana—formerly British Guiana—on South America's northeast coast, and they are the result of one young man's curiosity and initiative.

"I first heard from Balram Pertab in the spring of 1970," says Helen. "We'd had good productivity here in the summer of 1969 and then these bands began to come in to the banding office from Balram. He was netting the terns for his family's food, and also selling them in market. He continued to trap our birds in 1970 and 1971. He caught so many, in fact, that our government asked him to release the banded birds he trapped and not market them. He agreed to do this.

"In the winter of 1971 four of us visited Balram in Guyana to see how he caught the birds. He showed us the high beach near his village where birds could roost at night even with very high tides. He and his brother-in-law would go out with a gasoline lantern and a shrimp net; they'd dazzle the terns with the lamp, get close to them, and throw the net over them."

On the face of it, Balram Pertab is an unlikely prospect as a serious ornithological field worker, but in effect that's what he is, now. He has trapped terns from Great Gull Island in every month of the year. (Some of the young birds spend at least a full year and a half in Guyana after their first migration south before they head north again.) Balram makes careful records of his work and proudly keeps all the cards of thanks sent him by the Fish and Wildlife Service—one for every band he reports. He must be papering his walls with them by now; his trapped terns include banded birds from breeding grounds all over the Atlantic coast and on the Great Lakes.

William H. Drury, chief of the scientific staff of Massachusetts Audubon, wonders how many of Balram's friends in Whym are still capturing terns for food, and how many other South Americans are doing likewise along that vast stretch of coastline from the Isthmus of Panama south on both sides of the continent. How many terns are harvested each winter by growing, and impoverished, human populations? Thousands; tens of thousands, perhaps. Such harvests must have been going on for centuries on some scale, since long before Samuel Wyllys received his grant to Great Gull. But, as public health improves and population grows, is it now one reason for the drastic decline in numbers of New England terns that Bill Drury's colleague, Ian Nisbet, believes has become clearly evident? Are we farming terns partly so that people in coastal South America can eat and have something to sell at market? And if so, is that *bad?*

September 28–29: All the terns have left the island, bound for the unknown in more ways than one. Several times I thought I heard a faint *eee-ah* out over the Sound, and twice I saw small strings of far-distant terns beating their way southwestward. The bluefish have started to move; thus, so have the terns, so have the jaegers. Our "winter" cormorants, the greater or European cormorants, have begun to appear in numbers on the Sound, and the "summer" cormorants, the double-cresteds, will shortly string out in long Vs, like geese, for the south. Warblers and hawks trickled through Great Gull these two days, although the weather was southeasterly, not a spur to their migration.

We end as we began. It has been a weekend for tern-farming. We weeded purslane by the bushel from the roof of the great cavern back of the big gun emplacement, and we widened the "beaches" back of the Army's riprapping by cutting and tearing out yards of thick turf and long grass. A pump and motor had been brought out Friday night, and after Ron Franck had spent most of Saturday morning getting the thing to run and rigging long lengths of plastic hose to the pump, thousands of gallons of salt water were poured onto the newly opened beaches, to discourage the vegetation from taking hold again.

Next weekend there will be more of this. Then the summer's gear will be loaded on the *Sunbeam*—the rusty frames and canvas for the blinds, the limp cots and old sleeping bags, the mist nets and the banding equipment, the file cards and the notebooks, the two Coleman stoves and two dozen plastic jerricans and the canned goods and dried cereals and the tools and the telescopes—loaded on the *Sunbeam*, as it rises and falls in the chop beside the short dock. The lines will be cast off, Captain John will push the throttle ahead two-thirds, and at the end of the *Sunbeam*'s wake the ruins of Fort Michie will slowly shrink, become indistinct, and disappear.

Afterword

I am impressed by the work at Great Gull because of its commitment to life and the understanding of life. It seems to me that most of the human community is operating quite otherwise and has a collective death wish, which it is fulfilling (because of its collective *fear* of death) through inertia. Any parasitic organism—which man is, living off the earth—is totally successful only so long as it does not overpower its host. And we are overpowering our host, our only planet, wittingly but helplessly.

This began, of course, as the instinct for survival in an extremely adaptable, mobile, and dextrous creature: farm the earth, build shelters, create communities, connect the communities, exchange goods, live more safely and comfortably. The impulse simply got out of control, propelled by the admirable energies of the technologists. I am impressed by the argument that the cultural upcurve in human society has not kept pace with the technological upcurve. That is to say (for me), our spiritual relation to the planet is not yet nearly so sophisticated as our mechanical relation. Doubtless we have the theoretical capacity to achieve a balance, but as a species we do not have the desire and hence not the actual capacity. So the earth and our co-planetarians suffer under the weight of our population, our constructions, our territorializing, our wastes; and man puts his own existence in jeopardy.

Obviously, there must be reasons for the overpowering of the earth to this degree. Life is a discomfiting miracle; the existence of the planet is an enigma; a single lifetime is short; nature is awesome in its power and unpredictability. Does man try to overcome the earth because of his awareness that his span is merely a blip in fathomless time and space, and he himself a bit of chaff blown by infinite forces? That fearful awareness is forced on us every clear night of our lives, when we see the stars, and at every sunset and sunrise in which we look the sun in the face. Amidst all this space, we and our co-planetarians are the only life we know about. We explain it by different versions of God, who, being conveniently made in man's image—with various and sundry other comforting characteristics added—gives us a sense of power over the earth and the promise of life everlasting. But for men with the slightest sophistication, God—the *deus ex machina* indeed—is not a wholly satisfactory explanation. Men worship God for the miracle of life, while putting each other and their co-planetarians to death, striking repeatedly at the root and branch of their own collective existence; men worship God for the miracle of man, and honor war and hate and genocide in the observance as well as the breach. Despite the concept of God, the enigma remains, the contradictions multiply.

Our destruction of each other and our destruction of the planet will have the effect of putting an end to that. So we must feel we have reached our limit of understanding, or at least gone as far as our courage permits. And technology has not only given us new weapons to use against the earth and each other; it has vastly enlarged our awareness of our smallness and our isolation in space.

"The enterprise is exploration into God," says a character at the close of Christopher Fry's *A Sleep of Prisoners*, as the four protagonists—prisoners of life on earth—emerge trembling, terrified, from a dream passage through Nebuchadnezzar's fiery furnace. And that is the heart of the matter exactly: to survive as a species, man must dare an exploration into God, perhaps *beyond* what we have conceived of as God for two thousand years and more; an exploration, at any rate, into life.

So I observe, look for a *sign*, nibble at the edges of the unknown, and make my notes.